Concrete Repair

Also from Spon Press:

Management of Deteriorating
Concrete Structures
George Somerville
978-0-415-43545-1 (hbk)
978-0-203-93928-4 (ebk)

Testing of Concrete in Structures, 4th
edition
John Bungey et al.
978-0-415-26301-6 (hbk)
978-0-203-96514-6 (ebk)

Durability Design of Concrete in
Severe Environments
Odd E. Gjorv
978-0-415-41408-1 (hbk)
978-0-203-93141-7 (ebk)

Corrosion of Steel in Concrete, 2nd
edition
John Broomfield
978-0-415-33404-4 (hbk)
978-0-203-41460-6 (ebk)

Time-Dependent Behaviour of Concrete
Structures
Ian Gilbert et al.
978-0-415-49384-0 (hbk)
978-0-203-87939-9 (ebk)

Reynolds's Reinforced Concrete
Designer's Handbook
Anthony J. Threlfall et al.
978-0-419-25820-9 (hbk)
978-0-419-25830-8 (pbk)
978-0-203-08775-6 (ebk)

Multi-Scale Modeling of Structural
Concrete
Koichi Maekawa et al.
978-0-415-46554-0 (hbk)
978-0-203-92720-5 (ebk)

High-Strength Concrete
A practical guide
Mike Caldarone
978-0-415-40432-7 (hbk)
978-0-203-96249-7 (ebk

Concrete Repair
A practical guide

Edited by Michael G. Grantham

LONDON AND NEW YORK

First published 2011
by Taylor & Francis

Published 2017 by Routledge
2 Park Square, Milton Park, Abingdon, Oxfordshire OX14 4RN
711 Third Avenue, New York, NY 10017, USA

First issued in paperback 2017

Routledge is an imprint of the Taylor & Francis Group, an informa business

British Library Cataloguing in Publication Data
A catalogue record for this book is available from the British Library

Library of Congress Cataloging-in-Publication Data
Grantham, Mike.
Concrete repair : a practical guide / Michael Grantham.
       p. cm.
  Includes bibliographical references and index.
   1. Concrete construction--Maintenance and repair--Handbooks, manuals, etc.
  I. Title.
TA682.4.G69 2011   2010035082
624.1′8340288--dc22

ISBN 13: 978-1-138-07355-5 (pbk)
ISBN 13: 978-0-415-44734-8 (hbk)

Typeset in Sabon by
HWA Text and Data Management, London

Contents

Figures

Tables

Contributors

John P. Broomfield, DPhil. (Oxon), EurIng, FICorr, FIM3, FNACE, is an independent consulting engineer, specialising in the corrosion of steel in concrete and of atmospherically exposed steel framed buildings. He carried out some of the earliest impressed current cathodic protection installations outside North America. Recent projects include Selfridges Department Store, the Cutty Sark Historic Tea Clipper, Battersea Power Station, Coventry Cathedral and the University of East Anglia. He is convenor of CEN TC 219 WG2 on cathodic protection of steel in concrete. His book *Corrosion of Steel in Concrete* is now in its second edition.

John Drewett, BSc, CEng, MICE, MICorr is the Group Marketing Director for Concrete Repairs Ltd. He has worked for CRL since 1984 and has been responsible for the development of cathodic protection and composite strengthening systems for the UK construction market. He has represented CRL on many technical committees for the publication of standards and research programmes in the concrete repair sector.

Mike Eden is a Partner at Sandberg LLP and manages the Geomaterials Department. He is a Petrographer with more than 20 years' experience in the petrographic examination of a very wide range of civil engineering construction materials including concrete, grout, mortar, aggregates and crushed rock fill. He has published Research papers on delayed ettringite formation (DEF) in concrete, alkali-aggregate reaction (AAR) in concrete and the thaumasite form of sulphate (TSA) attack in concrete. He is author of the *Applied Petrography Group Code of Practice* for the Petrographic Examination of Concrete APG-SR2 and a contributor to Concrete Society publications and the Report of the Thaumasite Expert Group.

Michael G. Grantham has worked in the field of non-destructive testing and concrete repair for many years. He is a former Associate Director of STATS Ltd (now RSK Stats Ltd), a former Director of Technotrade Ltd, specialising in construction materials testing and is currently the Director of Concrete Solutions (GR Technologie Ltd.) and is organiser of the Concrete Solutions series of International Conferences on Concrete Repair. He is a member of Council of both the Concrete Society and the Institute of Concrete Technology. He is also a Visiting Professor at Queen's University, Belfast and a Consultant to Sandberg LLP. He is co-author of several books on concrete inspection and repair and Editor-in-Chief of the *Concrete Solutions Conference Proceedings*, currently published by Taylor and Francis.

Shaun A. Hurley, BSc PhD MRSC, was a Senior Materials Consultant with The Technology Division of VINCI Construction UK Ltd at the time of his retirement in December 2009. He joined (what was then) Taywood Engineering Ltd in 1987 after working for Cementation Research Ltd for 15 years. Concerned mainly with polymer-based materials, he has extensive experience in product formulation, materials testing, research, consultancy and technical service to support site operations. In addition to many contributions to the open literature, he has lectured widely on materials used in construction, while also serving on a number of technical and Standards committees.

Willy H. A. Peelen is a researcher in the group materials performance, risk and reliability at TNO. He has been involved in many national and international research and consultancy projects in the fields of corrosion, corrosion protection, monitoring and durability with applications in building structures, roads, bridges, tunnels and pipelines. He works on numerical modelling of these processes and leads the development of innovative monitoring systems for civil infrastructure together with stakeholders. He is a member of EFC working group 11, 'Concrete Reinforcement Corrosion'. Recently he has been involved in developing sustainable durability in civil infrastructure.

Rob B. Polder is a senior materials scientist at TNO, The Netherlands Organisation for Applied Scientific Research, in the group materials performance, risk and reliability; and a full professor of materials and durability at Delft University of Technology, faculty Civil Engineering and Geosciences. The main focus of his work is on corrosion of steel in concrete, from modelling and prediction to prevention and remediation, including electrochemical methods. He is or has been a member of many international and national research committees on durability of concrete. He was the chairman of European Action COST 534 New materials and systems for prestressed concrete structures.

Peter Robery is a Technical Director with consulting engineers Halcrow Group Ltd specialising in the investigation and repair of concrete defects and deterioration. A Fellow of the ICE and Past President and Fellow of the Concrete Society, Peter works extensively as an Expert Witness in dispute cases, with over 30 years of concrete design and construction experience, and has amassed over 60 papers, articles and book contributions. He is a Visiting Professor at the University of Leeds and Queens University Belfast and has represented BSI on CEN Committee TC104/SC8 and is responsible for the EN1504 standard on concrete repair and protection.

Paul Russell has worked for BASF for the past 21 years, and is responsible for composite strengthening. Paul has been involved with various committees and steering groups including the Concrete Society TR 55 and 57 reports, The Skills training working groups for composites for the DTi as well as SMM6. Paul is currently involved in the steering group BoF Heriot Watt Frp project, as well as being an active member of the NGCC (National Group for Composites in Construction). Paul has held training modules for students at Manchester, Leeds and Birmingham Universities giving them an insight into the use of composite materials in new build and refurbishment projects.

Ulrich Schneck, PhD, is a very experienced corrosion engineer, specialising in the corrosion behaviour of concrete structures and their remediation. He is the General Manager of CITec Concrete Improvement Technologies GmbH, a Body Member of DIN NA062-01-71, German Delegate to European Standards Committee TC219 WG2, TC219 WG5 and co-chair of the EFC WP11 Taskgroup on Corrosion Monitoring. He also lectures for EIPOS at TU, Dresden and is co-organiser of the forthcoming Concrete Solutions International Conference on Concrete Repair in Dresden, in 2011.

Graham Taylor, BSc and Fellow of the Institute of Concrete Technology, was for some time a consultant to the Sprayed Concrete Association. Whilst at the Cement and Concrete Association he was a catalyst for recognising sprayed concrete as a viable repair method. He is a former Executive Officer of the Institute of Concrete Technology.

G. P. Tilly, PhD, BSc, ACGI, FICE is Emeritus Director, Gifford and Partners, and a Visiting Professor at the University of Exeter. Dr Tilly is a civil engineer with some 40 years' experience in the design and management of both modern and historic bridges. At the former Road Research Laboratory he was Head of the Bridges Division and latterly an Assistant Director and Head of Structures. In 1992 he became Director of R and D with Gifford and Partners. He has written four books including *Conservation of Bridges* which was awarded the International Lee Nelson Book Award in 2003.

Jonathan G. M. Wood, BSc PhD FIStructE MICE CEng is Director of Structural Studies & Design Ltd. He is a specialist in forensic investigation of durability and structural failures and development of enhanced specifications for durability for tunnels, bridges and buildings. He was RAEng visiting Professor at Aston University 1996–2006. Keynote speaker at many international conferences on deterioration of construction materials and structural failures and on appropriate repairs and remedial works. He is also a Member of IStructE Task Groups on Appraisal of Existing Structures, Car Parks and Alkali Aggregate Reaction.

1 Understanding defects, testing and inspection

Michael G. Grantham

1.1 Introduction

Concrete, when it first came into use, was hailed as being long-lasting and largely maintenance free. The need for extensive repairs to concrete structures was hardly considered. And, with appropriate design of the concrete, with due recognition of the exposure conditions and a few simple rules in designing the mix and some simple pre-service testing, there is no reason why extended lifetimes for RC structures cannot be enjoyed. However, experience tells us that reinforced concrete is often not the maintenance-free material that some people expect and early failure sometimes ensues. It is worth exploring the reasons for this before proceeding.

Far too often, the basic design of the concrete and attention to detail in its placement are lacking. In the author's experience, many concrete mixes are not correctly designed and suffer a tendency to bleed and segregate, all of which renders the protection that should be afforded to the steel reinforcement compromised from the start.

Add to that incorrect placement of the steel, with inadequate cover and you have a recipe for early failure, whatever the exposure conditions. In severe conditions, with exposure to marine or de-icing salt, for example, failure can be very rapid indeed. Concrete Society Advice Note 17 (Roberts, 2006) lists some simple rules for good-quality surface finishes for cast-insitu concrete (see next section). Not surprisingly, those same rules will help to provide durable structures when combined with due attention to selection of an appropriate concrete and cover for the exposure conditions it will experience. The author once worked on the old Severn crossing, now replaced by the new bridge. The concrete in this bridge was of such high quality that it had not carbonated at all in 35 years and attempts to demonstrate a Windsor Probe device for strength estimation, caused the pin to simply bounce off, as the concrete was so hard!

There are other reasons why premature failure can occur: in the 1960s, for example, following pressure from the industry to improve formwork stripping times, the cement industry responded by grinding the cement finer and increasing the tricalcium silicate content (C3S) in the cement. Readymix producers were quick to spot that it was now possible to achieve the 28-day

Figure 1.1 Zero cover provided to a column in a reinforced concrete car park.

design strength with less cement (and therefore a higher water to cement ratio). Whereas, previously, concrete tended to gain in strength significantly after 28 days, with the newer cements that strength gain was much lower. The end result of this was concrete that carbonated rapidly, resulting in the onset of reinforcement corrosion much sooner than expected. There has also been a tendency for contractors to leave the procurement of the concrete to the buying department. Left with a free hand, they will choose the cheapest concrete possible – typically with about a 50 mm slump (S1 consistence). This concrete may be totally unsuitable to place in the works, where a higher workability may well be required, so the temptation to add some water on site to improve placeability is huge. The result, again, is a higher than anticipated water to cement ratio and lower durability. The correct procedure, of course, is to decide on an appropriate set of concretes for the different parts of a contract, with appropriate workability in each case. These can then be called off as required, and the temptation to add water avoided!

Designers can also improve the chances of a durable structure by avoiding placing drip details directly under a bar (or at least providing additional cover or protection in these areas). Here, the drip groove is placed underneath a horizontal bar, with typically 5–10 mm cover at the top of the groove! The author cannot count the number of times this simple error has been observed on structures during his career!

Another wonderful example of poor durability design is the use of so-called 'reconstituted stone' for window mullions and sills. These are made from a semi-dry mortar mix rammed into moulds and usually contain one or more steel rods for handling purposes. An alternative, sometimes found, comprises a rather poorly compacted concrete core, containing the steel rod, with a well-formed attractive stone-like, sandy coloured, mortar facing. Both types have a tendency to carbonate rapidly and corrosion of the steel then ensues with splitting and spalling of the units. Since these are often in use on high-rise structures and offices, the risk of falling material causing injury is high. Indeed, when tackled on this issue, one company, who shall remain nameless, advised the author that carbonation was no longer an issue, because they carbonate the units at the factory! He was of course referring to the other problem that reconstituted stone can suffer – flaking and crumbling of edges and corners, due to inadequate water and compaction in the mix. Carbonating the concrete hardens the surface and thus helps to avoid damaged edges, but compromises the durability of any units containing steel reinforcement, unless this is galvanised or otherwise protected.

Billions of pounds are spent each year on the repair of structures. In the US, alone, for example, annual repair costs are estimated at 18–21 billion USD (source: American Concrete Institute). Getting it right is therefore of critical importance.

In the remainder of this chapter we will explore some simple rules for avoiding problems in the first place and how non-destructive testing (NDT) and semi-destructive tests like coring and lab analysis can pay dividends in avoiding or diagnosing defects.

1.2 Get the concrete right

To ensure good-quality, well-finished and inherently durable concrete in the first place, a few simple steps need to be followed.

1   The correct quality of concrete, which has been designed to achieve an appropriate strength, durability for the exposure conditions it will experience and 'finishability'.
2   The use of the correct type and quality of form-face material and release agent suitable for the finish specified.
3   Workmanship, both in producing the formwork and mixing, placing, compacting and finishing the concrete.

The concrete itself should meet the criteria given in Table 1.1.

Table 1.1 Guidelines for achieving good surface finish (from Roberts, 2006)

1	Cement content	Minimum 350 kg/m^3
2	Sand content	Not more than twice the cement content
3	Total aggregate	Not more than six times the cement content
4	Coarse aggregate	For 20 mm max. size – ideally not more than 20% to pass a 10 mm sieve
5	Consistence	Not critical, but appropriate for good placement and compaction around the steel
6	Water/cement ratio	Normally 0.5 or less

And, of course, the steel reinforcement should be provided with appropriate cover for the exposure conditions. These seem simple enough, but it is remarkable how often these simple rules are not followed and the concrete is compromised from day one.

1.3 The role of non-destructive testing

1.3.1 Tests at the time of construction

NDT can be used initially at the time of construction to help to ensure that the structure is correctly built with the right materials and the right cover. While cube tests are being conducted, it is entirely possible to carry out some Schmidt hammer tests (Figure 1.2) on the side of the cube prior to loading and support these with some ultrasonic pulse velocity (UPV) measurements. In this way, a set of reference data for strength against Schmidt hammer and UPV can be built up. If there is any question regarding the quality of concrete in a site component, or maybe just randomly on site, these techniques can then be used to confirm the strength and quality of the concrete, as placed. Far too often, if there is a dispute regarding cube tests or cylinder tests, the engineer will call for core tests to be carried out. The relationship between core tests and cube tests is very complex, depending on the type of member, its curing, the type of concrete, orientation and a range of other factors. Core testing often results in posing more questions than it answers. It is possible using BS EN 13791 (BSI, 2007) and the recently published BS 6089 (BSI, 2010) which offers complementary guidance to BS EN 13791, to estimate the in-situ strength of the concrete in a component. This can be compared with the design strength and a decision can then be made on adequacy of the concrete. In the author's view, attempting to correlate the core and original cube tests, however, is so fraught with difficulty that it is not recommended. Table 1.2, reproduced from a Concrete Society Technical Report (Concrete Society, 2004), illustrates the problem. The data shows the variability between core and cube strength for the same concrete cast into different types of members, with different cement types, at different ages.

Figure 1.2 Schmidt hammer (courtesy Proceq UK Ltd).

Apart from establishing that the concrete is of the right quality and that it has been correctly placed, cover to the reinforcement is the next most important parameter. Figure 1.3 illustrates a modern covermeter. Engineers, in the author's experience, rarely know how to use covermeters and have no idea of the importance of bar size on the results and how misleading cover data can be obtained if the bar size is set incorrectly, or if lapped bars are encountered. These details are covered in the section on covermeters that follows. At the time of construction, it is entirely possible to check that the steel has the appropriate cover both before and after placing the concrete. This can be done through timber formwork (though not with steel formwork), simply deducting the timber thickness from the results. The covermeter only 'sees' the steel reinforcement and timber or concrete have no influence (though some aggregates, such as Lytag, can rarely interfere due to a property called 'magnetic viscosity' (Elcometer, 2010). Thus it is possible to ensure that the cover is correct at the point of construction, while the concrete has not yet hardened and while there is still time to intervene if a mistake has been made.

1.3.2 Understanding the problem

It is critically important to understand what has caused the deterioration in the structure and to fully understand the extent and reasons for it, prior to deciding on a repair strategy. Failure to do so can mean an inappropriate repair is chosen, which will inevitably fail more quickly than a correctly selected repair. For example, failure to understand that chloride contamination in the concrete has caused reinforcement corrosion, and simply attempting to patch repair it, causes early failure because new corrosion cells immediately form around the patch repair (Broomfield, 2007), resulting in further spalling

Table 1.2 Variation between core and cube strength in concrete (Concrete Society, 2004)

Cement	Mixes	Element	28 Days		42 Days		84 Days		365 Days	
			Min	Max	Min	Max	Min	Max	Min	Max
PC	Lower Strength Mixes	Block	0.75	0.95	0.80	0.95	0.85	1.05	0.95	1.10
		Slab	1.00	1.25	1.05	1.35	1.05	1.50	1.15	1.60
		Wall	0.85	0.95	0.85	1.00	0.85	1.05	0.90	1.10
	Higher Strength Mixes	Block	0.60	0.70	0.70	0.75	0.70	0.80	0.75	0.90
		Slab	0.85	1.00	0.95	1.10	0.95	1.15	1.05	1.25
		Wall	0.85	0.95	0.80	0.95	0.85	1.00	0.95	1.05
P/FA-B	Lower Strength Mixes	Block	0.85	1.10	0.85	1.20	0.95	1.35	1.45	1.65
		Slab	0.80	1.20	0.90	1.30	1.00	1.50	1.65	2.15
		Wall	0.80	1.05	0.85	1.10	1.05	1.30	1.50	1.60
	Higher Strength Mixes	Block	0.75	1.00	0.80	1.00	0.95	1.05	1.00	1.35
		Slab	0.75	0.95	1.00	1.15	1.15	1.25	1.20	1.70
		Wall	0.80	1.00	0.90	1.05	1.05	1.25	1.35	1.40
P/B	Lower Strength Mixes	Block	0.70	1.10	0.85	1.10	1.00	1.20	1.15	1.35
		Slab	0.65	1.15	0.85	1.30	1.20	1.40	1.35	1.70
		Wall	0.70	1.00	0.85	1.15	1.05	1.25	1.15	1.35
	Higher Strength Mixes	Block	0.80	1.05	0.95	1.20	0.90	1.25	1.00	1.55
		Slab	0.90	1.05	1.05	1.30	1.20	1.30	1.30	1.80
		Wall	0.70	1.00	0.85	1.10	1.00	1.20	1.05	1.25
PLC	Lower Strength Mixes	Block	0.80	0.95	0.85	1.00	0.95	1.10	1.00	1.15
		Slab	0.95	1.05	1.05	1.20	1.15	1.35	1.20	1.45
		Wall	0.85	1.00	0.90	1.05	0.95	1.00	1.10	1.20
	Higher Strength Mixes	Block	0.60	0.80	0.70	0.90	0.70	1.00	0.80	0.95
		Slab	0.80	0.90	0.95	1.00	1.00	1.05	1.05	1.20
		Wall	0.80	0.95	0.90	1.00	0.90	1.05	1.00	1.15

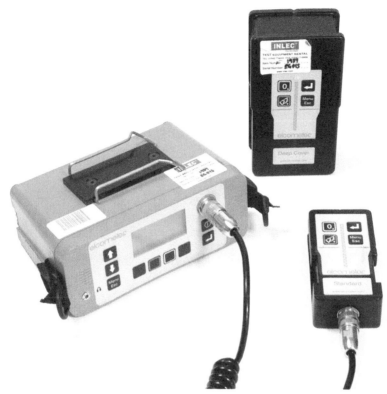

Figure 1.3 Elcometer 331S covermeter (courtesy Elcometer Ltd).

around the new repair. There are many other ways that misunderstanding the problem will result in a poor repair.

1.3.3 Recognising concrete defects

Structural failure

Actual structural failure, or even structural cracking, is only rarely encountered but it is important to differentiate between cracking from structural and other causes. Such an assessment should only be carried out by a structural engineer. Figure 1.4 illustrates structural cracks at a column/ beam connection. If examined carefully, the column behind shows no evidence of cracking, but in this case the beam is supported on a brick wall. The wall was not intended to be load bearing, but was clearly taking some of the stress off the column.

Figure 1.4 Structural cracks in a column due to overloading/inadequate reinforcement.

Figure 1.5 shows a car park with early signs of failure due to punching shear. This causes radial cracking around column positions as the deck flexes about the support. At later stages, an annular crack appears, which can indicate the onset of failure. Pipers Row car park in Wolverhampton (Figure 1.6) collapsed partly due to this phenomenon.

Corrosion and corrosion damage in concrete

The main causes of corrosion of steel in concrete are chloride attack and carbonation. Damage due to the latter is illustrated in Figure 1.7. These two mechanisms are unusual in that they do not attack the integrity of the concrete. Instead, aggressive chemical species pass through the pores in the concrete and attack or depassivate the passive oxide layer. Carbon dioxide and the chloride ions are very unusual in penetrating the concrete without significantly damaging it. The author has, however, come across clauses in repair bills requesting 'remove all weak, carbonated concrete'. Carbonated concrete is not necessarily weak: indeed carbonation hardens the concrete. It is only an issue where there is steel reinforcement within the carbonated zone. Of course the concrete might be carbonated because it is weak, then strength may be an issue for the structure.

The corrosion of steel reinforcement which is used in concrete is caused by electrochemical reactions. During corrosion, a part of the steel surface becomes anodic (corroding) and metal dissolution occurs, with small charged particles, called ions, going into solution. Further along the bar a

Figure 1.5 Cracks due to the onset of punching shear in a car park deck slab.

Figure 1.6 Pipers Row car park in Wolverhampton, after failure (photo courtesy of HSE).

Figure 1.7 Low cover and carbonation induced corrosion.

cathodic (non-corroding) region develops where oxygen gas and moisture which have diffused through the concrete to the steel surface react with electrons produced from the anodic reaction to produce hydroxyl ions.

$$2Fe \rightarrow 2Fe^{2+} + 4e^- \quad \text{anodic region}$$
$$O_2 + 2H_2O + 4e^- \rightarrow 4OH^- \quad \text{cathodic region}$$

The iron then reacts with the hydroxyl ions to form iron hydroxide and then, by further oxidation, iron oxide or rust (Figure 1.8).

Corrosion of the reinforcement will not occur if the pore solution of the concrete in which the steel is embedded remains sufficiently alkaline (assuming chlorides are absent and that stray electric currents are not present). In an alkaline state, a passive oxide layer forms over the steel reinforcement which prevents corrosion. However, carbon dioxide (and sulfur dioxide) in the air are acidic gases, and react with the concrete to cause alkalinity to be lost from the surface inwards: a gradual process known as carbonation. In good-quality concrete of low permeability, the rate of movement of the carbonation front is very slow, and provided the concrete has a reasonably low water to cement ratio, and sufficient concrete cover has been provided, the integrity of the reinforcement will last for the lifetime of the structure.

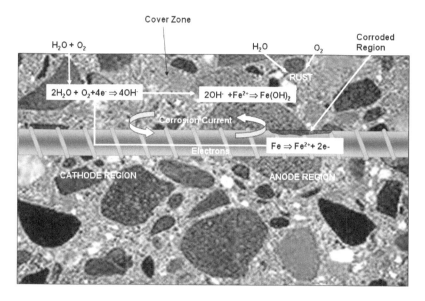

Figure 1.8 Schematic showing mechanism of corrosion in reinforced concrete.

The protection of the reinforcement therefore depends on the prevention of diffusion of carbon dioxide, and in some environments, such as bridges, chloride salts.

The type of corrosion which occurs varies. Carbonation-induced corrosion tends to affect large areas of the bar, causing a gradual loss of section over a relatively wide area. The corrosion problem is obvious before serious damage can be done because the concrete cover will spall. With chlorides, a different mechanism often occurs, causing very localised severe loss. An example of corrosion caused by both carbonation and chloride is given in Figure 1.9. This can occur without disruption of the cover concrete and almost total loss of section can occur before problems become apparent at the surface. Where pre-stressed steel is used, catastrophic failures have occurred with no prior warning, even on a structure which had been load tested shortly before the failure occurred!

Figure 1.9 illustrates corrosion caused by both carbonation and ingress of wind-blown marine salt to an office structure near to the sea. Interestingly, we had surveyed the same structure three years previously, and only splitting of the columns was apparent. In three years the damage had progressed to become a public safety matter, and was verging on causing structural problems.

Figure 1.9 Carbonation and chloride-induced damage on a seafront structure.

Carbonation

Carbon dioxide gas is an acidic gas, when dissolved in water, and reacts with the alkaline hydroxides in the concrete. Being a weak acid, the carbonic acid does not significantly attack the cement paste, but just neutralises the alkalis in the pore water, mainly forming calcium carbonate.

$$CO_2 + H_2O \rightarrow H_2CO_3$$
gas   water   carbonic
acid

$$H_2CO_3 + Ca(OH)_2 \rightarrow CaCO_3 + 2H_2O$$
carbonic pore   calcium   water
acid   solution   carbonate

There is a lot more calcium hydroxide in the concrete pores than can be dissolved in the pore water. This helps maintain the pH at a high level as the carbonation reaction occurs. However, eventually all the locally available calcium hydroxide reacts, precipitating the calcium carbonate and allowing the pH to fall to a level where steel will corrode. This is illustrated in Figure 1.10.

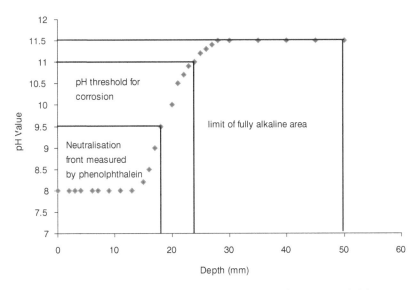

Figure 1.10 Graph showing variation of pH with depth (from Broomfield, 2007).

Carbonation-induced corrosion damage occurs most rapidly when there is little concrete cover over the reinforcing steel. Carbonation can occur even when the concrete cover depth to the reinforcing steel is high. This may be due to a very open pore structure where pores are well connected together and allow rapid CO_2 ingress. It may also happen when the alkaline 'reserves' in the pores are low. These problems occur when there is a low cement content, high water to cement ratio and poor curing of the concrete.

A simple rule for carbonation is that the depth of carbonation (in mm) should roughly equal the square root of the age in years. There are numerous models available for the estimation of carbonation depth, see (Broomfield, 2007).

Carbonation depth (mm) = k(age in years)$^{1/2}$

where k is a constant and roughly equal to 1 for good-quality concrete

However, the value of k depends on humidity, type of surface finish and a number of other factors.

At the carbonation front there is a sharp drop in alkalinity (Figure 1.10) although it can be seen that a small zone exists where corrosion is possible though phenolphthalein would remain pink. Once the pH drops, the passive oxide layer, created by the alkalinity, is no longer stable so corrosion can commence in the presence of sufficient air and moisture.

Micro-silica and other additives can block pores or reduce pore sizes, thus reducing carbonation. However, such materials can also use up some of the reservoir of calcium hydroxide and reduce the possible buffer to carbonation. Carbonation is common in old structures, poorly-built structures and reconstituted stone elements containing reinforcement that often have a low cement content and are very porous. Conversely, well-built structures such as bridges are often much less prone to carbonation (although unfortunately, historically, not to chloride ingress).

Carbonation is easy to detect and measure. A pH indicator, usually phenolphthalein in a solution of water and alcohol, will detect the change in pH across a freshly exposed concrete face. Phenolphthalein changes from colourless at low pH (carbonated zone) to pink at high pH (uncarbonated concrete). Measurements can be taken on concrete cores, fragments and by breaking the bridge between twin drilled holes used for chloride sampling. Care must be taken to prevent dust or water from contaminating the surface to be measured, but the test, with the indicator sprayed on to the surface, is cheap and simple. Petrographic examination can provide further information and is capable of detecting carbonation down cracks that phenolphthalein may not detect. A detailed report on petrographic examination is available from the Concrete Society (Concrete Society, 2010). As a cautionary note, phenolphthalein is now regarded as carcinogenic, and appropriate precautions to avoid skin contact and inhalation of spray must be taken.

Chlorides in concrete

Chloride salts, when present in reinforced concrete, can cause very severe corrosion of the steel reinforcement. Chlorides can originate from two main sources:

a   'Internal' chloride, i.e. chloride added to the concrete at the time of mixing. In this category, calcium chloride accelerating admixtures, contamination of aggregates and the use of sea water or other saline contaminated water are included.

b   'External' chloride, i.e. chloride ingressing into the concrete post-hardening. This category includes both de-icing salt as applied to many highway structures and marine salt, either directly from sea water in structures such as piers or in the form of air-borne salt spray in structures adjacent to the coast.

The effect of chloride salts depends to some extent on the method of addition. If the chloride is present at the time of mixing, the calcium aluminate (C3A) phase of the cement will react with the chloride to some extent, chemically binding it as calcium chloroaluminate. In this form, the chloride is insoluble in the pore fluid and is not available to take part in damaging corrosion reactions. The ability of the cement to complex the chloride is limited, however, and depends on the type of cement. Sulfate-resisting cement, for example, has a low C3A content and is therefore less able to complex the chlorides. In any case, experience suggests that if the chloride exceeds about 0.4% by mass of cement, the risk of corrosion increases. This does not automatically mean that concretes with chloride levels higher than this are likely to suffer severe reinforcement corrosion: this depends on the permeability of the concrete and on the depth of carbonation in relation to the cover provided to the steel reinforcement.

When the concrete carbonates, by reaction with atmospheric carbon dioxide, the bound chlorides are released. In effect this provides a higher concentration of soluble chloride immediately in front of the carbonation zone. Normal diffusion processes then cause the chloride to migrate into the concrete. This process, and normal transport of chlorides caused by water soaking into the concrete surface, is responsible for the effect sometimes observed where the chloride level is lower at the surface, but increases to a peak a short distance into the concrete (usually just in front of the carbonation zone). The increase in unbound chloride means that more is available to take part in corrosion reactions, so the combined effects of carbonation and chloride are worse than either effect alone.

Passivation of the steel reinforcement in concrete normally occurs due to a two-component system comprising a portlandite layer and a thin pH stabilised iron oxide/hydroxide film on the metal surface (Leek and Poole, 1990). When chloride ions are present, the passivity of the system is lost by dissolution of the portlandite layer, followed by debonding of the passive film. Physical processes operating inside the passive film may also contribute to its disruption.

The critical chloride content required to initiate corrosion depends on whether the chloride was present at the time of mixing, or has ingressed post-hardening, as discussed above. Clearly this also depends on the microclimate of the concrete (temperature and humidity) and also whether the concrete has carbonated. Typical contents are about 0.2% by mass of cement where chlorides are added at the time of mixing and slightly higher (0.4–0.5%) of cement for chloride ingressing post-hardening of cement. Good quality concrete can often show a remarkable tolerance for chloride without significant damage, however, at chloride contents up to about 1% by mass of cement (usually for chloride added at the time of mixing; reinforced concrete is much less tolerant of ingressed chloride).

When chlorides have ingressed from an external source particularly, in conditions of saturation and low oxygen availability, insidious pitting corrosion can occur, causing massive localised loss of cross-section. This can occur in the early stages without disruption of the overlying concrete.

Alkali silica reaction (ASR or AAR)

Alkali silica reaction (ASR) is the most common form of alkali aggregate reaction (AAR) and can occur in concretes made with aggregates containing reactive silica, as long as there is a sufficient supply of alkali (usually provided by the cement) and a supply of moisture.

The reaction product is a hygroscopic gel which takes up water and swells. This may create internal stresses sufficient to crack the concrete. If core samples are washed and wrapped in cling film, when taken, concrete suffering from ASR often develops dark, sweaty patches under the cling film surface, as gel oozes out of the concrete. This is a very good diagnostic indication that ASR is occurring, but does not necessarily indicate that damaging expansion either has or will result. This can only be shown by petrography and expansion testing.

One of the most frequently found aggregates in affected concrete is chert. This is a common constituent of many gravel aggregates, but a number of other geological types may be reactive, such as strained quartz in sands and some quartzites. Some Irish aggregates, notably greywackes, have been found to be susceptible to ASR. These tend to be quite slow reacting and damage can take 20–50 years to become serious. Greywackes have also caused problems in Wales.

Figure 1.11 illustrates cracking to the Beauharnois Dam on the St. Lawrence Seaway. The expansion caused by the ASR on this structure caused it to grow several cm in length and distorted the turbines providing hydro-electric power for the region.

Guidance on diagnosis of ASR is given in a BCA report (BCA, 1992). A photomicrograph of ASR in petrographic thin section is given in Figure 2.5 in Chapter 2.

Other rarer forms of reaction include alkali carbonate reaction and alkali silicate reaction. Discussion of these is beyond the scope of this chapter.

Freeze-thaw damage

Concrete of inadequate durability, if subjected to a wet environment and freezing, can be disrupted by freeze–thaw attack. Water enclosed in the pores of the wet concrete will expand on freezing and the high internal stresses so created can disrupt the surface. The effects are intensified by subsequent freeze–thaw action as minute cracks develop which, in turn, become filled with water. Again, concrete with a high water to cement ratio is especially vulnerable. Addition of an air-entraining agent to the concrete reduces the risk of frost damage by entraining minute, closely spaced air voids in the concrete, which provide a release mechanism for the expansive disruptive forces.

The problem is often characterised by parallel lines of cracking as freeze–thaw damage penetrates deeper into the concrete. Leaching of calcium hydroxide from the concrete also often occurs as water passes through the cracked concrete.

Figure 1.11 ASR-induced cracking on a dam in Canada.

Shrinkable aggregates

Some, mostly igneous, aggregates can contain inclusions of weathered material in the form of clay minerals. These minerals, in common with the clays encountered in the ground, swell in the presence of moisture and shrink as they dry out (Figure 1.13). They can cause excessive drying shrinkage of the concrete and can cause a random crack pattern not unlike that encountered with ASR. The problem was first identified in Scotland, where it is quite common, but has been observed in the North East of England, Hertfordshire, Wales and Cornwall. The cracking can pose potential structural problems but is more likely to cause loss of durability and is frequently associated with freeze–thaw damage. Figure 1.13 illustrates a severe example accompanied by pop outs where frost has attacked the porous aggregate.

Figure 1.12 Concrete with severe shrinkage cracking due to shrinkable aggregates.

Figure 1.13 Freeze thaw damage to a concrete slab – note the loose aggregate exposed as the cement matrix erodes (photo courtesy of the Concrete Society).

Sulfate attack

Concrete buried in soils or groundwater containing high levels of sulfate salts, particularly in the form of sodium, potassium or magnesium salts, may be subjected to sulfate attack under damp conditions. An expansive reaction occurs between the sulfates and the C3A phase of the hydrated cement to form calcium sulfoaluminate (ettringite) with consequent disruption to the matrix. As a guide, levels of sulfate above about 4% of cement (expressed as SO^3) may indicate the possibility of sulfate attack, provided sufficient moisture is present. Sulfate attack requires prolonged exposure to damp conditions. However, there has been recent concern with another form of sulfate attack, as follows

Thaumasite attack (TSA) – a form of sulfate attack

The industry became aware of this problem in 1998, when the foundations to a number of bridges on the M5 were found to be suffering from serious erosion and crumbling of the outer part of the concrete in the foundations (Figure 1.14). The problem was diagnosed as being due to an unusual form of sulfate attack, known as thaumasite attack.

For the problem to occur, a number of factors have to be present:

- A source of sulfate
- Water (usually plenty of moisture)
- A source of limestone (as aggregate, or filler, or possibly as fill or even carbonated groundwater)
- Low temperatures (<15°C)

Figure 1.14 Example of thaumasite/ettringite attack (courtesy of Halcrow plc).

The combination of these factors can cause an unusual reaction between the cement, the carbonate and the sulfate, to form *thaumasite*, a sulfate mineral. The effect is to cause serious damage and softening of the exposed outer surface of the concrete (assuming an external source of sulfate). Figure 1.14 shows one of the affected foundations.

It should be noted that sulfate-resisting cement has not proved to be any more resistant than normal portland cement (CEM1) in resisting this type of attack.

An international conference to discuss thaumasite attack was organised by the Building Research Establishment (BRE). The proceedings of the conference are available (BRE, 2002).

Acid attack

Acid attack is typified by 'raised' aggregate, or in extreme cases disintegration. It can sometimes be difficult to diagnose since the acid is neutralised by the cement paste and may be washed away. The most common sources are spillage from acid tanks, acidic groundwater and oxidation of sewage effluents. The latter can cause quite dramatic damage to concrete unless appropriate precautions are taken.

Other contaminants

In addition to those specifically mentioned above, many potentially corrosive substances may come into contact with reinforced concrete. The extent of their effect may depend on the type of cement used.

Examples of such materials are certain alkalis, beer and wine (carbonic acid, lactic acid and acetic acid), vegetable and fish oils, milk (lactic acid), lime, sugar, sulphides, manure and silage.

Fire damage

Damage and repair of concrete due to fire is summarised in Technical Report No.68 (Concrete Society, 2008) of the Concrete Society. Three principal types of alteration are usually responsible.

1   Cracking and microcracking in the surface zone: This is usually sub-parallel to the external surface and leads to flaking and breaking away of surface layers. Cracks also commonly develop along aggregate surfaces – presumably reflecting the differences in coefficient of linear expansion between cement paste and aggregate. Larger cracks can occur, particularly where reinforcement is affected by the increase in temperature.
2   Alteration of the phases in aggregate and paste: The main changes occurring in aggregate and paste relate to oxidation and dehydration. Loss of moisture can be rapid and probably influences crack development. The paste generally changes colour and various colour zones can

develop. A change from buff or cream to pink tends to occur at about 300°C and from pink to whitish grey at about 600°C. Certain types of aggregate also show these colour changes which can sometimes be seen within individual aggregate particles. The change from a normal to light paste colour to pink is most marked. It occurs in some limestones and some siliceous rocks – particularly certain flints and chert. It can also be found in the feldspars of some granites and in various other rock types. It is likely that the temperature at which the colour changes occur varies somewhat from concrete to concrete and if accurate temperature profiles are required, some calibrating experiments need to be carried out.

3   Dehydration of the cement hydrates: This can take place within the concrete at temperatures a little above 100°C. It is often possible to detect a broad zone of slightly porous light buff paste which represents the dehydrated zone between 100 and 300°C. It can be important, in reinforced or pre-stressed concrete, to establish the maximum depth of the 100°C isotherm.

CHANGES IN FIRE-DAMAGED CONCRETE

<300°C	Boundary cracking alone
250–300°C	Aggregate colour changes from pink to red
300°C	Paste develops a brown or pinkish colour
300–500°C	Serious cracking in paste
400–450°C	Portlandite converts to lime
500°C	Change to anisotropic paste
500–600°C	Paste changes from red or purple to grey
573°C	Quartz gives a rapid expansion resulting from a phase change from alpha to beta quartz
600–750°C	Limestone particles become chalky white
900°C	Carbonates start to shrink
950–1000°C	Paste changes from grey to buff

CHANGES IN AGGREGATE

250–300°C	Aggregate colour changes from pink to red
573°C	Quartz gives a rapid expansion resulting from a phase change from alpha to beta quartz
600–750°C	Limestone particles become chalky white
900°C	Carbonates start to shrink

CHANGES IN THE CEMENT PASTE

300°C	Paste develops a brown or pinkish colour
400–450°C	Portlandite converts to lime
500–600°C	Paste changes from red or purple to grey
950–1000°C	Paste changes from grey to buff

Figure 1.15 Poor compaction around the base of a pile.

CRACKING

< 300°C	Boundary cracking alone
300–500°C	Serious cracking in paste
500°C	Change to anisotropic paste

The depth of colour change is often a good guide to the overall depth of damage. This may often be no more than a few mm for short-duration fires, even if high temperatures are reached at the surface. As a general rule, if no spalling has occurred, no damage is likely to have occurred to the steel. (This does not apply to pre-stressed or post-tensioned concrete, which requires specialist assessment.) A structural engineer should always be involved in assessment of any repair, including fire damage.

Poor-quality construction

During construction, lack of attention to proper quality control can produce concrete which may be inferior in both durability and strength to that assumed by the designer. Particular factors in this respect are compaction, curing conditions, low cement content, incorrect aggregate grading, incorrect water/cement ratio and inadequate cover to reinforcement.

Plastic cracking

Cracks due to this phenomenon appear within the first two hours after placing and are of two distinct types:

1   Plastic settlement cracks: Typically found in columns, deep beams or walls. The problem tends to occur with high water/cement ratio concretes which have suffered from bleeding. The concrete literally 'hangs up' on the steel slumping between it, with cracks forming over the line of the steel. Caught early enough, re-vibration of the concrete can repair the damage while the concrete is still plastic. Figure 1.16 shows a core taken through a plastic settlement crack in a car park. The damage that the easy pathway for chlorides has caused can easily be seen.

2   Plastic shrinkage cracks: More common in flat exposed slabs. This can occur anywhere where the rate of loss of moisture due to evaporation exceeds the rate of bleeding. Not surprisingly, it is more of a problem in

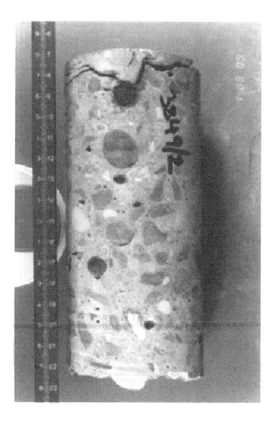

Figure 1.16 Plastic settlement cracks in a core.

Note: Neither of these should be confused with drying shrinkage cracks which only occur after a considerable time (though they may start as cracks from another cause).

hot, dry climates, but can easily occur on hot days in flat slabs especially where inadequate attention to protection and curing has been given.

Thermal cracking and delayed ettringite formation

Thermal cracking of concrete can occur in large pours. Typically concrete can gain in temperature about 14°C per 100 kg of cement in a cubic metre of concrete. In large pours this sets up a thermal gradient, with the outer part of the concrete cooling more rapidly than the core. This puts the outer skin in tension, and small cracks form. With the addition of subsequent drying shrinkage, the cracks can become quite large (Figure 1.17).

During the hydration process, it is quite normal for ettringite (calcium sulfoaluminate) to form inside the concrete. This mineral is normally associated with sulfate attack, but in the context of a setting concrete is quite normal. Any expansion resulting from its formation is taken up in the still-plastic concrete. However, if the temperature of the concrete exceeds about 70°C then formation of the ettringite can be delayed until after the concrete has hardened. In this situation, if a source of moisture is present, the concrete can suffer from quite severe expansion and cracking. Cracking due to DEF (Figures 1.18 and 1.19) can be mistaken for that due to ASR and is similar in nature (but not cause). There is a correlation with the alkali content of the cement, too. The higher the alkali content, the lower the temperature at which DEF can occur (Grantham et al., 1999).

Figure 1.17 Thermal cracks in the wall of a water tank.

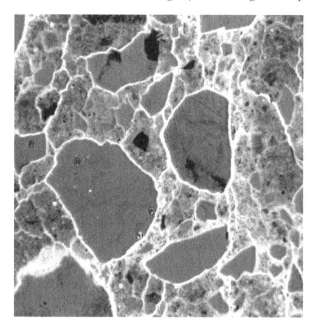

Figure 1.18 Cracking in the cement paste due to delayed ettringite formation. In this case, the cement paste expands, rather than the aggregate crack.

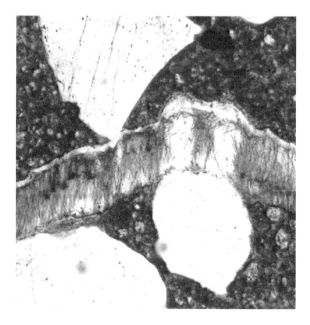

Figure 1.19 Ettringite crystals lining a crack.

Other types of cracking

For more detailed work on non-structural cracks in concrete, see Concrete Society Technical Report No. 22 'Non-Structural Cracks in Concrete' (Concrete Society, 2010).

1.4 NDT and laboratory methods

Any investigation can conveniently be split into two stages:

Stage 1: An initial survey to identify the cause of the problems.

Stage 2: An extension of the stage 1 survey, perhaps using a limited number of techniques to identify the extent of the defects revealed by stage 1.

The advantages of such an approach are clear. In the stage 1 survey, work can be carried out on selected areas showing typical defects but choosing these, as far as possible, from areas with simple access, i.e. ground level, roof level, from balconies, etc. Occasionally, a lightweight scaffold tower or an electrically powered hydraulic lift can be used to advantage. One or more areas apparently free from defect would also be examined in this initial survey as it is frequently found that, by comparing good areas with bad, the reason for the problems emerges by simple comparison. An excellent approach to stage 1 is to view the structure after heavy rain and to identify the weakest details from the design drawings. Where water sheds and where the weakest details are should most certainly be included in this initial survey.

In stage 2, once the defects have been identified, it is often necessary to quantify the extent of the problems. This may be as simple as carrying out a covermeter survey over the whole structure, where low cover has been identified as the problem, to the application of one or more of the other techniques described below.

1.4.1 Visual survey

After collecting as much background data as possible, any testing regime should begin with a thorough visual survey of the structure. This may conveniently be recorded on a developed elevation giving particular attention to the following defects:

- Cracks or crazing
- Spalling
- Corrosion of steel and rust staining
- Hollow surfaces
- Honeycombing due to poor compaction or grout loss
- Varying colour or texture
- Areas in which remedial finishing work had already been carried out
- External contamination or surface deposits
- Wet or damp surfaces.

Throughout the course of any investigation, colour photographs should be taken of points of particular interest.

1.4.2 Covermeter survey

Adequate cover to the steel reinforcement in a structure is important to ensure that the steel is maintained at a sufficient depth into the concrete so as to be well away from the effects of carbonation or from aggressive chemicals. However, excessively deep cover has its own problems; crack widths may be increased and the lever-arm decreased.

All covermeters are electromagnetic in operation. Electric currents in a coil winding in the search head generate a magnetic field which propagates through the concrete and will interact with any buried metal present, such as reinforcing steel. The interaction will be due to either or both of two physical properties of the steel: its magnetic permeability and its electrical conductivity. The interaction causes a secondary magnetic field to propagate back to the head where it is detected by a second coil or, in some instruments, by modifying the primary field. The signal received will increase with increasing bar size and decrease with increasing bar distance (cover). By making certain assumptions about the bar, and specifically by assuming that only one bar is present within the primary magnetic field, the instrument can be calibrated to convert signal strength to distance and hence to indicate the depth of cover.

If there is more than one bar (or even scaffolding) within the range of the primary field the instrument will receive a greater signal and indicate a shallower cover than the true cover. The skilled operator will always carefully map out the position and orientation of the steel, breaking out some steel, to ensure that accurate results are obtained. Some covermeters estimate bar size and can correct for size, so giving more accurate results if lapped bars are encountered.

Some manufacturers claim that the size of the reinforcing bar may be determined by the use of spacer blocks and some inbuilt mathematical processing, or with some of the more sophisticated machines by internal data processing only. Such methods work satisfactorily only where a single bar is present within the range of the search head, but can be reasonably accurate with some of the more modern devices. For best results, however, a breakout is always more trustworthy.

British Standard 1881 : Part 204 : 1988 (BSI, 1988) requires that when measuring cover to a single bar under laboratory conditions, the error in indicated cover should be no more than ±5% or 2 mm, whichever is the greater. For site conditions, an average accuracy of ±5 mm or 15% is suggested as being realistic in the British Standard. Recent developments in covermeters are now improving on this, with covermeters capable of showing better than 8% with an average of better than 2% over a wide range of bar sizes, lapped bars, etc. A survey of a car park carried out by the author, where cover to reinforcement was in contention, showed results on average within ±1% with a worst error of 3 mm, over a range of bar sizes, including lapped bars (calibrated against numerous breakouts). However, the cheapest covermeters are unlikely to achieve such an accuracy. Figure 1.20 illustrates one of the more expensive and competent covermeters.

The standard also lists a number of extraneous factors which are potential sources of error. Those concerned with magnetic effects from the aggregates or the concrete matrix, and those due to variations in cross-sectional shape of the bars should not affect the modern covermeter, but care must always be taken when dealing with multiple bars and the effects of adjacent steel such as window frames, crash barriers or scaffolding as mentioned earlier.

The Hilti Ferroscan Covermeter now provides an image of the reinforcement under the concrete surface (Figure 1.21) and a simulated 2D diagram of the reinforcement can also be obtained using the Profometer 5+ with a scanlog accessory, at significantly lower cost, but with less information.

1.4.3 Ultrasonic pulse velocity measurement (Pundit)

Introduction

The velocity of ultrasonic pulses travelling in a solid material depends on the density and elastic properties of that material. In concrete, the pulse velocity is related to its elastic stiffness so that measurement of ultrasonic pulse velocity can often be used to indicate concrete quality as well as to determine its elastic properties. Pulse-echo type surveys have only recently become available as a technique, which is known as ultrasonic tomography. Such surveys provide data not unlike that of radar and can be visualised by computer interpretation to give a 2D or 3D plot of features within the concrete (Lorenzi et al., 2010, Pinto et al., 2010). Figure 1.22 illustrates the PUNDIT 7 UPV equipment, available from Proceq.

UPV can be used to detect areas of concrete with defects (honeycombing, voidage, or simply high w/c ratio) as these will show a lower pulse velocity than sound areas. By appropriate calibration (and testing cube or cylinder samples ultrasonically prior to compression testing is a good way to achieve this) concrete can easily be tested to show whether it is of consistent quality throughout a member, or from member to member, provided these are of similar age. Considerable use of this technique was made in the evaluation of structures made with high-alumina cement, for example, to follow any loss in strength with time. Similarly, there is no reason why such a tool cannot be used for routine quality assessment on new structures to ensure consistent quality between pours and between different parts of a structure. The technique is detailed in British Standard BS EN 12504-4:(BSI, 2004).

Figure 1.20 Proceq Profometer 5 covermeter (courtesy Proceq Ltd).

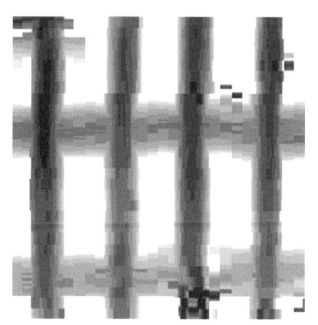

Figure 1.21 Hilti Ferroscan image of steel reinforcement in a slab.

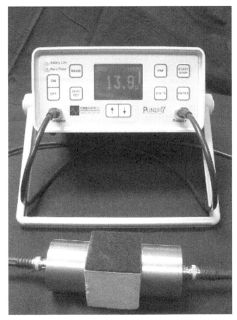

Figure 1.22 PUNDIT 7 UPV instrument (courtesy Proceq Ltd).

Applications

The pulse velocity method of testing may be applied to the testing of plain, reinforced and pre-stressed concrete whether it is pre-cast or cast in situ.

Measurement

The measurement of pulse velocity may be used to determine:

a   the homogeneity of the concrete,
b   the presence of voids, cracks or other imperfections,
c   changes in the concrete which may occur with time (i.e. due to the cement hydration) or through the action of fire, frost or chemical attack,
d   the quality of the concrete in relation to specified standard requirements, which generally refer to its strength.

Accuracy

In most of the applications it is necessary to measure the pulse velocity to a high degree of accuracy since relatively small changes in pulse velocity usually reflect relatively large changes in the condition of the concrete. For this reason it is important that care be taken to obtain the highest possible accuracy of both the transit time and the path length measurements since the pulse velocity measurement depends on both of these.

It is desirable to measure pulse velocity to within an accuracy of ±2% which allows a tolerance in the separate measurements of path length and transit time of only a little more than ±1%.

When such accuracy of path length measurement is difficult or impossible, an estimate of the limits of accuracy of the actual measurements should be recorded with the results so that the reliability of the pulse velocity measurements can be assessed.

Estimation of strength

Concrete quality is generally assessed by measuring its cube (or cylinder) crushing strength. It has been found that there is no simple correlation between cube strength and pulse velocity but the correlation is affected by:

a　type of aggregate,
b　aggregate/cement ratio,
c　age of concrete size and grading of aggregate,
d　curing conditions.

In practice, if pulse velocity results are to be expressed as equivalent cube strengths, it is preferable to calibrate the particular concrete used by making a series of test specimens with materials and mix proportions the same as the specified concrete but with a range of strengths. The pulse velocity is measured for each specimen which is then tested to failure by crushing. Figure 1.23 illustrates a typical UPV/Strength relationship.

The range of strength may be obtained either by varying the age of the concrete at test or by introducing a range of water/cement ratios. In practice, on real contracts, tests on cubes prior to compression testing will enable a calibration curve to be obtained.

If strength monitoring with time is to be carried out, the calibration curve is best obtained by varying the age but a check on quality at a particular age would require the correlation to be obtained by varying the water/cement ratio.

For more information on the use of ultrasound in the examination of concrete, see Bungey et al., 2006 and the Pundit Reference Manual available from Proceq Ltd.

1.4.4 Chemical tests

Chemical analysis of concrete can provide extremely useful information regarding the cause or causes of failure of concrete. The tests most frequently carried out are listed below:

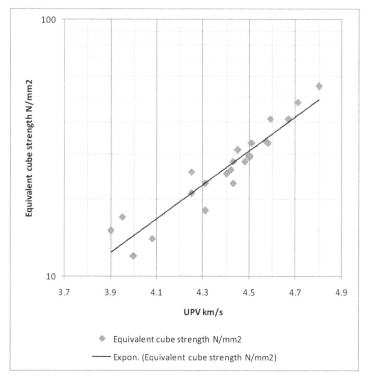

Figure 1.23 Typical relationship between cube strength and UPV.

1.	Chloride content	4.	Sulfate content
2.	Cement content	5.	Type of cement
3.	Depth of carbonation	6.	Alkali content

In the following, an explanation of the reason why each parameter is important is given, followed by an explanation of the test itself.

Chloride content test methods

The generally accepted method of test for chloride in hardened concrete is described in BS 1881 : Part 124: 1988 (BSI, 1988). The test involves crushing a sample of the concrete to a fine dust, extracting the chloride with hot dilute nitric acid and then adding silver nitrate solution to precipitate any chloride present. Ammonium thiocyanate solution is then titrated against the remaining silver and the amount of chloride determined from the difference between the added silver nitrate and that remaining after precipitating the chloride. Faster and more precise methods based on ion-selective electrodes are now available. One of the more modern automated analysers is illustrated in Figure 1.24. Automated testing is described in a paper from the Structural Faults and Repair Conference (Grantham, 1993)

Cement content

It is a fundamental requirement of good-quality concrete that it contains an adequate cement content, or more precisely, a sufficiently low water/cement ratio, to provide adequate durability for the intended exposure conditions. In the absence of chemical admixtures, a certain amount of water is required to provide an adequate workability; essentially to simply lubricate the aggregate particles and the cement. To achieve the desired water/cement ratio, the amount of cement required is therefore automatically defined. This can be altered only by changing the physical properties of the aggregate, or by the addition of a water-reducing admixture.

If the cement content is too low (i.e. the water/cement ratio too high), the concrete will be attacked by the weather and be liable to freeze–thaw damage and the effects of carbonation. If the cement content is too high, heat of hydration can cause thermal cracking in large pours, the risk of shrinkage increases (because of the higher water content) making curing doubly important, and, if a high-alkali cement is used, the risk of ASR increases with susceptible aggregates and DEF is more likely to occur.

TEST METHODS

The test to determine the cement content of concrete is given in BS 1881 : Part 124 : 1988 (BSI, 1988). It requires the crushed concrete to be extracted with dilute acid and dilute alkali solution to remove the cement. The extract

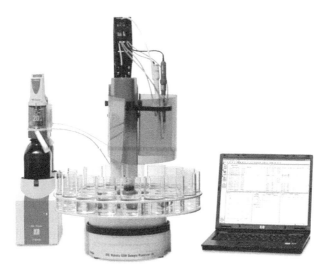

Figure 1.24 An automated analyser (courtesy Thermo Fisher).

is then analysed for soluble silica and calcium oxide, being the two major components (expressed as oxides) of Portland cement. The cement content is determined by simple proportion from the two parameters. Where soluble components from the aggregate interfere by contributing to the calcium content (e.g. if a limestone aggregate is present) then the silica value would be used for the cement content determination. Conversely, if the silica value was inflated by some soluble component other than the cement, the lime value would be used, provided the analyst was confident that this was unaffected by soluble components from the aggregate. In practice, it is normal to analyse control samples of the aggregate, where these are available, to avoid these problems. With control samples, an accuracy of better than ± 25 kg/m^3 is readily achievable.

Where cement replacement materials such as pfa (pulverised fuel ash) and ggbs (ground granulated blastfurnace slag) are present, the situation is more complex. Nevertheless, accurate results can often be obtained using total analyses by, for example, X-ray fluorescence methods and applying a simultaneous equations approach (Grantham, 1994) or by using petrographic methods combined with scanning electron microscopy (SEM). The topic is to be revisited in an update of Concrete Society Technical Report No. 32 (Concrete Society, 1989) by a Concrete Society working party.

Sulfate content

Exposure of concretes made with Portland cement to sulfate salts can cause damage due to an expansive reaction between the tricalcium aluminate phase of the cement and the sulfate salt to form crystals of ettringite. Given adequate space to form, the ettringite forms needle-like crystals, but in confined space causes an expansive reaction as the amorphous product develops. With good-quality concrete, significant sulfate attack is relatively rare, and research work suggests that concrete made with a reasonable cement content (at least 330 kg/m^3) and a reasonably low water/cement ratio, is attacked only very slowly. However, the most damaging salts are the more soluble sulfates based on magnesium or sodium sulfates. Calcium sulfate (gypsum) is only sparingly soluble and is less likely to cause damage. The rate of damage is also dependent on the rate of replenishment of the sulfate salts and hence on groundwater movement.

TEST METHODS

Sulfate is usually determined by the method given in BS 1881 : Part 124 : 1988 (BSI, 1988). This involves an acid extraction and precipitation of the sulfate as barium sulfate with barium chloride solution. The resulting barium sulfate is filtered and weighed to determine sulfate gravimetrically.

Methods based upon ion selective electrodes and ion chromatography have also been employed. Petrography is usually the best initial indicator of a sulfate attack problem.

High-alumina cement (HAC)

HAC achieved some notoriety during the 1970s following the collapse of several buildings in which it had been used (Building Research Establishment, 1974). This was due to a conversion of the cement from one crystalline form into another, weaker, form. At normal temperatures, the hydration of HAC results in the formation of hydrated calcium monoaluminate (CAH10). Smaller amounts of C2AH8 and hydrous alumina are also formed. However, these hydrated calcium aluminates are metastable and can, at higher temperatures and in the presence of moisture, change to give the stable hydrated calcium aluminate C3AH6. This phenomenon is known as 'conversion', and the amount of the change occurring, 'the degree of conversion'.

At normal temperatures, conversion may take many years but at temperatures in excess of 40°C a considerable amount of conversion can occur within a few months. Structures built with HAC are likely to have fully converted now and petrography and X-ray diffraction analysis are considered to be the best means of examination, with the degree of conversion test having fallen into disuse. Petrography should be accompanied by some means of strength evaluation and/or load testing.

Conversion results in a loss of strength, increased porosity and reduced resistance to chemical attack. There has been increasing concern regarding carbonation of high-alumina cement concrete. Following conversion, the increased porosity may permit rapid carbonation of the concrete, removing alkaline protection to the steel reinforcement, which may then suffer from corrosion.

TEST METHODS

A test was devised by the Building Research Station to show whether HAC was likely to be present in a concrete (BRE, 1974). It essentially tests for a significant content of soluble aluminium in solution, following extraction with dilute sodium hydroxide solution.

1.5 Petrographic examination (see also Chapter 2)

Preliminary examination

The samples are examined with the binocular microscope as received and their dimensions and main features are recorded. The features observed include the following:

a   The presence and position of reinforcement
b   The extent to which reinforcement is corroded
c   The nature of the external surfaces of the concrete
d   The features and distribution of macro and fine cracks
e   The distribution and size range and type of the aggregate

f   The type and condition of the cement paste
g   Any evidence of deleterious processes affecting the concrete.

Polished surfaces

A plate is cut, where possible, from each sample. This is typically about 20 mm thick and usually provides as large a section of the sample as is possible. The plate is polished to give a high quality surface that can be examined with a high quality binocular microscope or even with the petrological microscope if necessary. The polished plate is used to assess the following:

- The size, shape and distribution of coarse and fine aggregate.
- The coherence, colour and porosity of the cement paste.
- The distribution, size, shape and content of voids.
- The composition of the concrete in terms of the volume proportions of coarse aggregate, fine aggregate, paste and void.
- The distribution of fine cracks and microcracks. Often the surface is stained with a penetrative dye, so that these cracks can be seen. Microcrack frequency is measured along lines of traverse across the surface.
- The relative abundance of rock types in the coarse aggregate.

Thin sections

A thin section is prepared for each sample as appropriate. The section is usually made from a plate cut at right angles to the external surface of the concrete, so that the outer 70 mm or so of the concrete are included in the section. Sometimes it is more appropriate to make the section from inner parts of the concrete. This might be appropriate where specific problems are being investigated, for example. The section normally measures about 50 × 70 mm.

In manufacturing the thin section, a plate some 10 mm thick is cut from the sample. This is impregnated with a penetrative resin containing a yellow fluorescent dye. The resin penetrates into cracks, microcracks and capillary pores in the sample. One side of the impregnated plate is then polished and the plate is mounted on to a glass slide. The surplus sample is then removed and the plate is ground and polished to give a final thickness of between 20 and 30 micrometres. At all stages the cutting and grinding is carried out using an oil-based coolant in order to prevent further hydrating of the cement and excessive heating of the section. The thin section is covered and then examined with a high quality petrological photomicroscope (Figure 1.25). The thin section supplies the following types of information:

- Details of the rock types present in the coarse and fine aggregate and in particular structures seen within those rocks.

- Details of the aggregate properties are measured, such as the degree of strain in quartz.
- The size, distribution and abundance of phases in the cement paste are assessed including, for example, the occurrence of calcium hydroxide and the amount of residual unhydrated clinker.
- The presence of cement replacement phases such as slag or PFA can usually be recognised (though the amount of these phases cannot be judged accurately). The presence of high-alumina cement can be detected and the type of cement clinker can often be assessed.
- Any products of processes of deterioration of either the cement paste or the aggregate can be recognised.

Broken surfaces

After the specially prepared surfaces and sections are completed, the remainder of the core is examined with the binocular microscope. In particular, the pieces are broken to produce fresh surfaces. These surfaces allow the contents of voids to be studied and the nature of aggregate surfaces or crack surfaces to be investigated.

Composition

The composition of the sample is measured using either the polished slice or the thin section, depending on the size of the sample and on details of the aggregate type and paste. The thin section is preferable, for example where

Figure 1.25 Gel-filled cracks in concrete suffering from ASR, thin section.

large quantities of dust are present. The volume proportions are found by the method of point counting using a mechanical stage. The amount of coarse aggregate can also be assessed by this method if a distinction can be made between coarse and fine aggregate. The results obtained usually represent the sample reasonably, but may not represent the concrete.

The amount of individual rock types present in the aggregate as a whole are assessed and the saturated density of the sample is measured by the method of immersion in water using vacuum impregnation to ensure saturation. From this information and the volume proportions, the weight fractions of aggregate, cement and water can be calculated.

Water/cement ratio

The hydration processes of cement paste vary significantly with the original water/cement ratio. Concretes with a low water/cement ratio tend to leave substantial quantities of unhydrated cement clinker and to develop only limited amounts of coarsely crystalline calcium hydroxide. In particular, the extent to which calcium hydroxide is separated into layers on aggregate surfaces and occurs in voids and on void surfaces varies with the original water/cement ratio. The number and proportion of unhydrated cement clinker particles varies inversely with the original water/cement ratio. Comparison with standard concretes made with known water/cement ratios visually and by measurement allows the water/cement ratio of the cement paste to be assessed directly. The standard error attached to the estimation of water/cement ratio by this means is considered to be approximately ±0.03.

A Concrete Society Technical Report has recently been published giving full details of what can be achieved using this technique (Concrete Society, 2010).

1.6 Schmidt hammer (rebound hammer)

The rebound principle for concrete testing for surface hardness is widely accepted. The most popular equipment, the Schmidt rebound hammer, has been in use world-wide for many years. Recommendations for the use of the rebound method are given in BS EN 12504-2 (BSI, 2001) and ASTM C805 (ASTM, 2008).

Rebound test equipment and operation

The Swiss engineer Ernst Schmidt first developed a practicable rebound test hammer in the late 1940s and modern versions are based on this. A spring-controlled hammer mass slides on a plunger within a tubular housing. The plunger retracts against a spring when pressed against the concrete surface and this spring is automatically released when fully tensioned, causing the hammer mass to impact against the concrete through the plunger. When the spring-controlled mass rebounds, it takes with it a rider which slides along a scale and is visible through a small window in the side of the casing. The rider can be held in position on the scale by depressing the locking button. The equipment is very simple to use and may be operated either horizontally or vertically either upwards or downwards.

The plunger is pressed strongly and steadily against the concrete at right angles to its surface, until the spring-loaded mass is triggered from its locked position. After the impact, the scale index is read while the hammer is still in the test position. Alternatively, the locking button can be pressed to enable the reading to be retained or results can automatically be recorded by an attached paper recorder. The scale reading is known as the rebound number, and is an arbitrary measure since it depends on the energy stored in the given spring and on the mass used. This version of the equipment is most commonly used, and is most suitable for concrete in the 20–60 N/mm² strength range. Electronic digital reading versions of the equipment are available.

Procedure

The equipment is very sensitive to local variations in the concrete, especially to aggregate particles near to the surface and it is therefore necessary to take 12 readings in the area of interest and to average the results obtained. The surface to be measured should be smooth, clean and dry but if it is required to take measurements on trowelled surfaces, the surface can be smoothed using the carborundum stone provided with the instrument. This stone is *not* a calibration device and must not be used to receive an impact from the hammer as the stone may shatter.

Theory, calibration and interpretation

The test is based on the principle that the rebound of an elastic mass depends on the hardness of the surface upon which it impinges and in this case will provide information about a surface layer of the concrete defined as no more than 30 mm deep. The results give a measure of the relative hardness of this zone, and this cannot be directly related to any other property of the concrete. Energy is lost on impact due to localised crushing of the concrete and internal friction within the body of the concrete, and it is the latter, which is a function of the elastic properties of the concrete constituents, that makes theoretical evaluation of test results extremely difficult. Many factors influence results but must all be considered if rebound number is to be empirically related to strength.

Factors influencing test results

Results are significantly influenced by all the following factors:

a   Mix characteristics
     Cement type

    Cement content
    Coarse aggregate type
b   Member characteristics
    Mass
    Compaction
    Surface type
    Age, rate of hardening and curing type
    Surface carbonation
    Moisture condition
    Stress state and temperature.

Since each of these factors may affect the readings obtained, any attempts to compare or estimate concrete strength will be valid only if they are all standardised for the concrete under test and for the calibration specimens. These influences have different magnitudes. Hammer orientation will also influence measured values although correction factors can be used to allow for this effect and the Proceq Silver Schmidt has a built-in correction system for orientation.

For a detailed review of factors influencing the evaluation of surface hardness by the Schmidt hammer method see 'Testing of Concrete in Structures' (Bungey et al., 2006). One of the most important factors is surface carbonation. Concrete exposed to the atmosphere will normally form a hard carbonated skin, whose thickness will depend upon the exposure conditions and age. It may exceed 20 mm for old concrete although it is unlikely to be significant at ages of less than three months. The depth of carbonation can easily be determined by the phenolphthalein spray method. In extreme cases it is known that the overestimate of strength from this cause may be up to 50% and is thus of great importance. When significant carbonation is known to exist, the surface layer ceases to be representative of the concrete within an element

Calibration

Clearly, the influences of the variables described above are so great that it is very unlikely that a general calibration curve relating rebound number to strength, as provided by the equipment manufacturers, will be of any practical value. The same applies to the use of computer data processing to give strength predictions based on results from the electronic rebound hammer, unless the conversions are based on case-specific data. Strength calibration must be based on the particular mix under investigation, and the mould surface, curing and age of laboratory specimens should correspond as closely as possible to the in-place concrete. It is essential that correct functioning of the rebound hammer is checked regularly using a standard steel anvil of known mass. This is necessary because wear may change the spring and internal friction characteristics of the equipment. Calibrations prepared for one hammer will also not necessarily apply to another. It is probable that very few rebound hammers used for in-situ testing are in fact regularly checked against a standard anvil, and the reliability of results may suffer as a consequence.

The importance of specimen mass has been mentioned above; it is essential that test specimens are either securely clamped in a heavy testing machine or supported upon an even solid floor. Cubes or cylinders of at least 100 mm should be used, and a minimum restraining load of 15% of the specimen strength has been suggested for cylinders, and BS EN 12504-2 recommends not less than 7 N/mm^2 for cubes tested with a type N hammer. Typically the relationship between rebound number and restraining load is such that once a sufficient load has been reached the rebound number remains reasonably constant.

It is well established that the crushing strength of a cube tested wet is likely to be about 10% lower than the strength of a corresponding cube tested dry. Since rebound measurements should be taken on a dry surface, it is recommended that wet cured cubes be dried in the laboratory atmosphere for 24 hours before test, and it is therefore to be expected that they will yield higher strengths than if tested wet in the standard manner.

Note that a calibration *must* be performed for each concrete that is to be tested, before the relationship between cube strength and rebound number can be established. The values in Figure 1.26 accord well with the author's experience, however.

Other near-to-surface strength tests

These include the Windsor Probe, the BRE Internal Fracture Tester, various break-off devices and the CAPO and Lok tests used extensively in Scandinavia and increasingly in the UK to establish concrete maturity.

Radar profiling

Over the past thirty years or so there has been an increasing usage of sub-surface impulse radar to investigate civil engineering problems and, in particular, concrete structures. Electromagnetic waves, typically in the frequency range 500 MHz to 1.5 GHz, will propagate through solids, with the speed and attenuation of the signal influenced by the electrical properties of the solid materials. The dominant physical properties are the electrical permittivity which determines the signal velocity, and the electrical conductivity which determines the signal attenuation. Reflections and refractions of the radar wave will occur at interfaces between different materials and the signal returning to the surface antenna can be interpreted to provide an evaluation of the properties and geometry of sub-surface features.

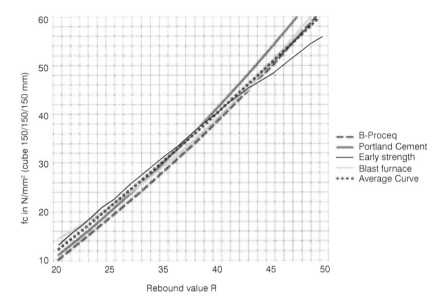

Figure 1.26 Typical rebound number/strength relationship (courtesy of Proceq Ltd).

1.7 Radar systems

There are three fundamentally different approaches to using radar to investigate concrete structures.

a   Frequency modulation: in which the frequency of the transmitted radar signal is continuously swept between pre-defined limits. The return signal is mixed with the currently transmitted signal to give a difference frequency, depending upon the time delay and hence depth of the reflective interface. This system has seen limited use to date on relatively thin walls.
b   Synthetic pulse radar: in which the frequency of the transmitted radar signal is varied over a series of discontinuous steps. The amplitude and phase of the return signal is analysed and a 'time domain synthetic pulse' is produced. This approach has been used to some extent in the field and also in laboratory transmission line studies to determine the electrical properties of concrete at different radar frequencies.
3   Impulse radar: in which a series of discrete sinusoidal pulses within a specified broad frequency band are transmitted into the concrete, typically with a repetition rate of 50 kHz. The transmitted signal is often found to comprise three peaks, with a well-defined nominal centre frequency.

Impulse radar systems have gained the greatest acceptance for field use and most commercially obtainable systems are of this type. The power output of the transmitted radar signal is very low, and no special safety precautions are needed. However, in the UK a Department of Trade and Industry radiocomms licence is required to permit use of investigative radar equipment.

Radar equipment

Impulse radar equipment comprises a pulse generator connected to a transmitting antenna. This is commonly of a bow-tie configuration, which is held in contact with the concrete and produces a divergent beam with a degree of spatial polarisation. A centre frequency antenna of 1–1.5 GHz is often used in the investigation of relatively small concrete elements, up to 500 mm thick, while a 500 MHz antenna may be more appropriate for deeper investigations. However, a lower frequency loses resolution of detail despite the improved penetration.

An alternative to using surface-contact antennae is to use focused beam horn antenna with an air gap of about 300 mm between the horn and the concrete surface. These systems have been used in the USA and Canada to survey bridge decks from a vehicle moving at speeds of up to 50 km/hr, principally to detect corrosion-induced delamination of the concrete slab. Operational details are provided in ASTM D4748 (ASTM, 2006).

Structural applications and limitations

In addition to the assessment of concrete bridge decks, radar has been used to detect a variety of features buried within concrete, ranging from reinforcing bars and voids to murder victims.

The range of principal reported structural applications is summarised in Table 1.3. Interpretation of radar results to identify and evaluate the dimensions of sub-surface features is not always straightforward. The radar picture obtained often does not resemble the form of the embedded features. Circular reflective sections such as metal pipes or reinforcing bars, for example, present a complex hyperbolic pattern due to the diverging nature of the beam. The use of signal processing can simplify the image but interpretation is still complex. Evaluating the depth of a feature of interest necessitates a foreknowledge of the speed at which radar waves will travel through concrete. This is principally determined by the relative permittivity of the concrete, which in turn is determined predominantly by the moisture content. Figure 1.27 shows radar in use on a concrete investigation.

Because of the difficulties of interpretation, surveys are normally conducted by specialists who rely on practical experience and have a knowledge of the limitations of the technique in practical situations. For example, features such as voids can be particularly difficult to detect if located very deep or beneath a layer of closely spaced reinforcing steel. Neural networks or

Table 1.3 Structural applications of radar

Reliability:

Greatest → Least

Determine major construction features

Assess element thickness

Locate reinforcing bars

Locate moisture

    Locate voids, honeycombing, cracking

        Locate chlorides

            Size reinforcing bars

                Size voids

                    Estimate chloride concentrations

                        Locate reinforcement corrosion

Figure 1.27 Radar in use on a concrete investigation (Courtesy Fugro Aperio Ltd.).

'artificial intelligence' has been used to help with interpretation of complex radar traces.

Some modern equipment has the ability to present data as 2D or even 3D plots showing an indication of the detected features, which are easier to interpret than the parabolic-type reflections normally encountered in radar plots.

Radar reflects most strongly off metallic objects or from the interface between two materials with widely differing permittivities. An air-filled void in dry concrete, which does not differ very strongly in permittivity from the concrete itself, can therefore be difficult to detect, especially if the void is small. The same void filled with water, however, would be much more easily detected.

Water strongly attenuates a radar signal and using a 1 GHz antenna, typical practical penetrations of around 500 mm have been achieved for dry concrete, and 300 mm for water-saturated concrete. If the water is contaminated with salt, penetration is likely to be smaller still.

Radar Antenna: TX: Transmitting Antenna \ RX: Receiving Antenna

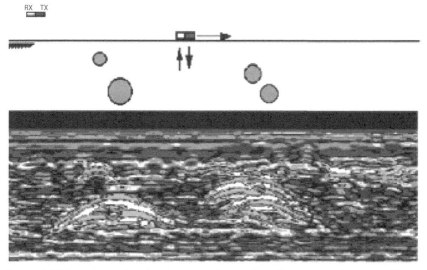

Figure 1.28 Impulse radar showing trace from rebars in concrete (courtesy Fugro-Aperio Ltd).

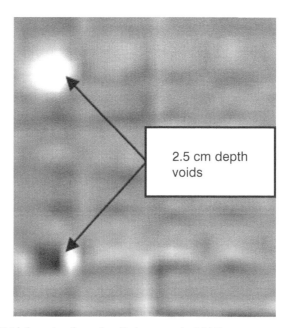

Figure 1.29 Void detection by radar (Roberts et al., 2010).

1.8 Infra-red thermography

This technique uses infra-red photographs taken from a structure which has been heated, as it cools. The heating is normally performed by the sun in daytime and the photographs are best recorded in the evening as the structure cools. Infra-red thermography offers many potential advantages over other physical methods for the detection of delamination in bridge decks. Areas of sound and unsound concrete will exhibit different thermal characteristics and thus have different surface temperatures as the structure cools. Delaminated areas, for example, will have a different temperature gradient compared with sound areas. Water-saturated concrete will appear quite different to dry concrete. The temperature differences are small but are readily visible on the highly sensitive modern infra-red cameras, such as the RAZ-IR™. The images can be downloaded to a PC for further analysis.

The surface of the structure needs to be viewed from a reasonable distance and so cannot be recorded whilst, for example, standing on a deck. Some success has been achieved working from a tall truck at a height of 20 m, provided that the temperature differences were at least 2°C. Working from an aircraft or helicopter avoids the need for lane closures but has not consistently shown good results. Figures 1.30 and 1.31 demonstrate the usefulness of IR photography in revealing delamination in a silo tower.

Holt and Eales (1987) have also described the successful use of thermography to evaluate effects in highway pavements with an infra-red scanner and coupled real-time video scanner mounted on a 5 m high mast attached to a van.

The vehicle is driven at up to 15 mph and images are matched by computer. Procedures for infra-red thermography in the investigation of bridge deck delamination are given in ASTM D4788 (ASTM, 2007).

Hidden voids or ducts can also sometimes be detected, and techniques have been developed to detect reinforcing bars which have been heated by electrical induction. More recent development of 12-bit equipment has improved the sensitivity to within ±0.1°C. This has enabled high-definition imaging and accurate temperature measurement on buildings. The smallest detectable area is reported to be 200 × 200 mm.

Infra-red thermography can also be used to reduce heat losses at hot spots by identifying missing thermal insulation.

1.9 Testing for reinforcement corrosion

1.9.1 Half cell potential testing

Steel embedded in good-quality concrete is protected by the high alkalinity pore water which, in the presence of oxygen, passivates the steel. The loss of alkalinity due to carbonation of the concrete or the penetration of chloride ions (arising from either marine or de-icing salts, or in some cases present

Figure 1.30 Delamination in a post-tensioned silo.

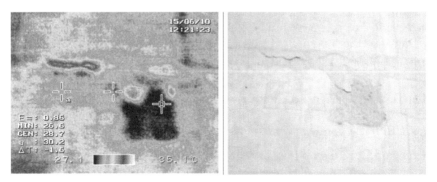

Figure 1.31 IR photograph (left) of delaminated area of the silo in Figure 1.30. Right is a normal photograph.

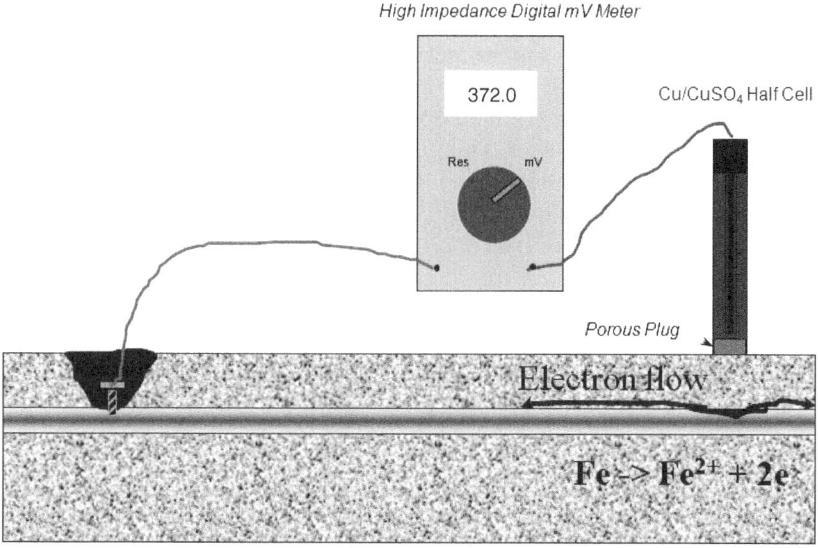

High Impedance Digital mV Meter

372.0

Cu/CuSO$_4$ Half Cell

Res   mV

Porous Plug

Electron flow

$Fe \rightarrow Fe^{2+} + 2e^-$

Figure 1.32 Half cell potential testing – schematic.

in situ from the use of a calcium chloride additive) can destroy the passive film. In the presence of oxygen and humidity in the concrete, corrosion of the steel starts. A characteristic feature for the corrosion of steel in concrete is the development of macrocells – the co-existence of passive and corroding areas on the same reinforcing bar with the corroding area as the anode and the passive surface as the cathode. The voltage of such a cell can reach as high as 0.5 V or more, especially where chloride ions are present. The resulting current flow (which is directly proportional to the mass lost by the steel) is determined by the electrical resistance of the concrete and the anodic and cathodic reaction resistance.

The current flow in the concrete is accompanied by an electrical field which can be measured at the concrete surface, resulting in equipotential lines that allow the location of the most corroding zones at the most negative values. This is the basis of potential mapping, the principal electrochemical technique applied to the routine inspection of reinforced concrete structures (Broomfield, 2007).

The use of the technique is described in an American Standard, ASTM C876-09, Standard Test Method for Half Cell Potentials of Reinforcing Steel in Concrete (ASTM, 2009). The standard has changed considerably from its earlier incarnation in 1980 and now contains numerous caveats and more guidance on factors influencing half cell potential values.

In use, a reference connection is made by exposing a reinforcing bar and connecting by brazing or by drilling a small hole and inserting a self-tapping screw into the bar (CAUTION – this method is not suitable for prestressed or post-tensioned steel reinforcement, where only a mechanical connection such as a crocodile clip can be permitted). The reference connection is connected to the positive terminal of a high impedance digital millivoltmeter. A reference cell such as a copper/copper sulfate half cell is attached to the negative terminal of the meter. This cell has a porous plug at one end, which permits an electrical connection. A grid is marked on the surface of the concrete at suitable intervals, usually between 0.5 to 1 m spacings in a square grid pattern. The surface of the concrete to be tested is wetted with a wetting agent at each node to be measured on the grid intersections. The meter is touched to the concrete surface and the millivoltmeter reading noted. The cell is then removed and touched again and a second reading taken. The readings should be within 10 mV of each other. Caution should be exercised if readings drift – stable values should be obtained.

Factors affecting the potential field

When surface potentials are taken, they are measured remote from the reinforcement due to the concrete cover. The potentials measured are therefore affected by the ohmic potential drop in the concrete. Several factors have a significant effect on the potentials measured:

CONCRETE COVER DEPTH

With increasing concrete cover, the potential values at the concrete surface may indicate an average of local active and passive steel, making the location of corroding anodic areas more difficult.

CONCRETE RESISTIVITY

The concrete humidity and the presence of ions in the pore solution affect the electrical resistivity of the concrete. The resistivity may change both across the structure and with time as the local moisture and salt content vary. This may create an error of ±50 mV in the measured potentials (John et al., 1987).

HIGHLY RESISTIVE SURFACE LAYERS

The macrocell currents tend to avoid highly resistive concrete. The measured potentials at the surface become more positive and corroding areas may be undetected.

POLARISATION EFFECTS

Steel in concrete structures immersed in water or buried in the earth often have a very negative potential due to restricted oxygen access (Popovics et al., 1983). In the transition region of the structure (splash zone or above ground), negative potentials can be measured due to galvanic coupling with immersed rebars. These negative potentials are not related to corrosion of the reinforcement. Corrosion rate measurements can be useful in assessing the significance of these high potential values, as can resistivity measurements.

Results and interpretation

According to the ASTM method, corrosion can only be identified with 90% certainty at potentials more negative than –350 mV. Experience has shown, however, that passive structures tend to show values more positive than –200 mV and often positive potentials. Potentials more negative than –200 mV may be an indicator of the onset of corrosion. The patterns formed by the contours can often be a better guide in these cases.

In any case, the technique should never be used in isolation, but should be coupled with measurement of the chloride content of the concrete and its variation with depth and also the cover to the steel and the depth of carbonation. The ASTM document also suggests a potential difference technique where differences in potential are plotted over relatively small areas to locate corroding steel. This can be helpful in dry weather, for example, when corrosion cells become less active and may not always follow the ASTM guideline values. However, peaks of potential at corroding areas may well still exist. It has to be remembered, when evaluating results, that the technique measures what is happening *on the day*. Considerable variation in numeric values can be encountered over different seasons and weather conditions.

Figure 1.33 shows a survey of a car park deck slab using a 1 m grid. The corroding areas can be readily identified.

1.9.2 Resistivity

The electrical resistivity is an indication of the amount of moisture in the pores, and the size and tortuosity of the pore system. Resistivity is strongly affected by concrete quality, i.e. cement content, water/cement ratio, curing and additives used.

Equipment and use

The main device in use is the four-probe resistivity meter. These have been modified from soils applications and are used by pushing pins directly onto the concrete with moisture or gels to enhance the electrical contact. Millard et al. (1991) described two versions of the equipment. Some variations use drilled-in probes or a simpler, less accurate two-probe system. Further information can be found in Concrete Society Technical Report No 60 (Concrete Society, 2004).

Resistivity $= 2 \pi a V/I$ (Ωcm)

where: R is the resistance by the 'IR drop' from a pulse between a surface electrode and the rebar network measured by a half cell reference electrode

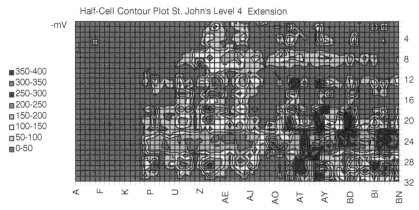

Figure 1.33 Half cell potential plot of a car park deck on a 1 m grid.

a   is the electrode spacing
I   is the current passed between outer probes (amps)
V   is the potential measured between inner probes (volts).

Interpretation

Interpretation is empirical. The following interpretations of resistivity measurements have been cited when referring to depassivated steel:

> 20 kΩcm   low corrosion rate
10–20 kΩcm low to moderate corrosion rate
5–10 kΩcm   high corrosion rate
< 5 kΩcm   very high corrosion rate.

Limitations

The resistivity measurement is a useful additional measurement to aid in identifying problem areas or confirming concerns about poor-quality concrete. Readings can only be considered when used with other measurements, such as chloride content, depth of carbonation, cover to reinforcement and half cell potential.

It should be understood that the resistivity does not tell you how fast the corrosion is or how much corrosion there is, it simply indicates the capacity of the concrete to support a corrosion cell *if one exists*. For example, a concrete showing a low half cell potential but also a low resistivity is not actively corroding, but could do if the conditions for corrosion existed. Conversely, a concrete showing a high half cell potential and a high resistivity is unlikely to show a significant rate of corrosion.

There are devices which can actually measure corrosion rate directly (Broomfield, 2007) such as the Gecor 8, although these have seen only limited use in the field.

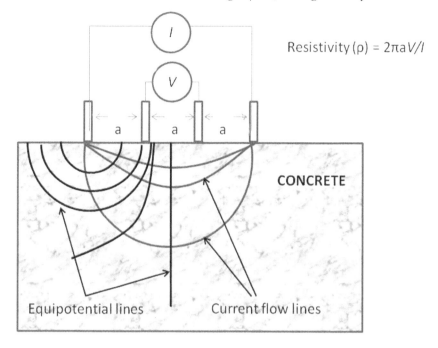

Resistivity $(\rho) = 2\pi a V/I$

Figure 1.34 Schematic showing typical 4-probe resistivity measurement.

1.10 Summary

Reinforced concrete can suffer from a significant number of defects, although good-quality, properly designed, well-mixed and placed concrete is normally durable for many years. It is critical, before approaching the repair of a structure, to understand what has caused the problem and to design a repair appropriate to that problem. Failure to do so can often result in a very short-lived repair. This chapter has attempted to introduce something of the range of different types of defect that can occur and some of the non-destructive and semi-destructive methods available to determine the causes of such problems.

People engaging in concrete repair should beware of attempting to identify the cause of a problem just from its appearance. Laboratory testing and in particular petrographic examination are almost always essential as part of the diagnostic process.

Once the problem is understood, an appropriate repair system can be designed.

References

ASTM D4748–06 'Standard Test Method for Determining the Thickness of Bound Pavement Layers Using Short-Pulse Radar,' ASTM, USA, 2006.

ASTM D4788–03 'Standard Test Method for Detecting Delaminations in Bridge Decks Using Infrared Thermography'. ASTM, USA, 2007.

ASTM C805/C805M – 08 'Standard Test Method for Rebound Number of Hardened Concrete.' ASTM, USA, 2008.

ASTM C876–09 'Standard Test Method for Half-Cell Potentials of Uncoated Reinforcing Steel in Concrete'. ASTM, USA, 2009.

British Cement Association, *The Diagnosis of Alkali Silica Reaction, Report of a Working Party*. 1992.

BRE Information Sheet IS15/74 'A Rapid Chemical Test for High Alumina Cement.' BRE, 1974.

Broomfield, J. *Corrosion of Steel in Concrete – Understanding, Investigation and Repair*. pp 120–122, 2nd edn, Taylor and Francis, 2007.

BS 1881 : 124:1988 'Methods for the Chemical Analysis of Hardened Concrete.' BSI, London, UK.

BS 1881:204:1988 'Recommendations on the Use of Electromagnetic Cover Measuring Devices'. BSI, London, UK.

BS 6089:2010 'Assessment of in-situ compressive strength in structures and precast concrete components.' Complementary guidance to that given in BS EN 13791, BSI, London.

BS EN 12504-2:2001. 'Testing concrete in structures. Non-destructive testing. Determination of rebound number.' BSI, London.

BS EN 12504-4:2004 'Testing concrete. Determination of ultrasonic pulse velocity.' BSI, London, 2004.

BS EN 13791:2007 'Assessment of in-situ compressive strength in structures and precast concrete components', BSI, London.

Building Research & Information, 'Beam Failure – Conversion of High Alumina Cement.' Volume 2, Issue 4 July 1974 , pp. 235–240.

Building Research Establishment, *Proceedings of the First International Conference on Thaumasite in Cementitious Materials*, BRE, 2002.

Bungey, J., Millard, S. and Grantham, M. *Testing of Concrete in Structures*. 4th edn, Taylor and Francis, 2006.

Concrete Society Technical Report No. 32, *Analysis of hardened concrete: a guide to test procedures and interpretation of results*, The Concrete Society, 1989.

Concrete Society Technical Report No. 60, *Electrochemical Tests for Reinforced Concrete*. Joint Concrete Society/Institute of Corrosion Report, Concrete Society, 2004.

Concrete Society, *In situ Concrete Strength – An investigation into the relationship between core strength and standard cube strength*. Concrete Society, 2004.

Concrete Society, *Assessment, Design and Repair of Fire-damaged Concrete Structures*, The Concrete Society, 2008.

Concrete Society Technical Report No. 22, *Non-Structural Cracks in Concrete*, The Concrete Society, 2010.

Concrete Society Technical Report No. 71, *Concrete Petrography*, The Concrete Society, UK, April 2010.

Elcometer Ltd, Concrete and Civil Engineering, Application and Technical Notes at http://www.elcometer.com/international%20index%20pages/international/Concrete/technical%20notes/A10.htm (accessed July 2010).

Grantham, M. G., 'An Automated Method for the Analysis of Chloride in Hardened Concrete.' *Proceedings of the 1993 Conference on Structural Faults and Repair.* Edinburgh, June 1993. ECS Publications.

Grantham, M. G. 'Determination of Slag and PFA in Hardened Concrete – The Method of Last Resort Revisited,' *Determination of the Chemical and Mineral Admixture Content of Hardened Concrete*, ASTM STP 1253, Steven H. Kosmatka and Ara A. Jeknavorian, eds, American Society for Testing and Materials, Philadelphia, 1994.

Grantham, M., Gray, M. and Eden, M. 'Delayed Ettringite Formation in Foundation Bases – A Case Study'. *Proceedings of the 8th International Structural Faults and Repair Conference* 1999. Engineering Technics Press.

Holt, F. B. and Eales, J. W. 'Non-destructive Evaluation of Pavements'. *Concrete International*. 9, No. 6, June 1987, pp.41–45.

John, D. G., Eden, D. A., Dawson, J. L. and Langford, P. E., *Proceedings of the Conference on Corrosion 1987*, San Francisco CA, 9.13.3 1987, Paper 136.

Leek, D. S. and Poole, A. B. 'The Breakdown of the Passive Film on High Yield Mild Steel by Chloride Ions.' *Proceedings of the 1990 SCI Conference on Corrosion of Reinforcement in Concrete*. Elsevier, 1990.

Lorenzi, A. et al. 'Strategies for monitoring concrete structures by means of ultrasonic pulse velocity.' *Proceedings, Structural Faults and Repair 2010*, Edinburgh.

Millard, S. G., Harrison, J. A. and Gowers, K. R. 'Practical Measurement of Concrete Resistivity.' *British Journal of NDT*. Vol. 33, No. 2. pp 59–63. February 1991.

Pinto, R. et al. 'Detection of concrete internal flaws using ultrasound direct method.' *Proceedings, Structural Faults and Repair 2010*, Edinburgh.

Popovics, S., Simeonov, Y. Bozhinov, G. and Barovsky, N. in *Corrosion of reinforcement in Concrete Construction*, ed. A. P. Crane, Society of Chemical Industry, London, 1983, pp 193–222.

Roberts, D., 'Concrete Advice Note 17, Achieving good quality as struck in situ concrete surface finishes.' December 2006, The Concrete Society, Camberley, UK.

Roberts, R., Corcorann, K. and Schutz, A. 'Insulated concrete form defect detection using ground penetrating radar.' *Proceedings of Structural Faults and Repair 2010*, Edinburgh.

2 The petrographic examination of concrete and concrete repairs

Mike Eden

2.1 Introduction

The most commonly employed methods of testing concrete for concrete repairs such as measuring compressive strength or cement content or pull-off testing for bond strength, leave many unanswered questions when it comes to assessing the condition of the concrete in a structure, or the overall effectiveness of a concrete repair. For example, the bond strengths measured by pull-off testing may be low but it is often not possible to determine why this is so without further testing. The measured cement content may be unexpectedly low, but it is not possible to determine without further testing whether or not the repair material contains a cement replacement material such as PFA that might affect the cement content measurement.

The key questions that can be answered by petrographic examination in evaluating a concrete structure in need of repair are:

- What is the current condition of the concrete?
- Is the concrete undergoing on-going deterioration?
- What depth of concrete needs to be removed before repair materials can be applied?

For evaluating an existing concrete repair, the following additional questions can be answered from petrographic examination:

- In what condition is the bond between the repair and the substrate?
- Has the repair material deteriorated?

This chapter introduces petrographic examination as a cost-effective method capable of providing information about many different aspects of the condition of a concrete repair, the substrate and the bond between the repair and the substrate. It describes how cementitious repair materials may be identified petrographically and describes the diagnosis of problems with concrete repair durability.

2.1.1 Removing concrete samples from concrete and concrete repairs

Core drilling

Concrete samples obtained by diamond core drilling are ideal for petrographic examination as core drilling minimises the possibility of introducing damage to the sample during sampling. Water-lubricated diamond core drilling of concrete repairs exerts very little stress on the sample or the bond between the repair and the substrate. Because many types of concrete repair materials lack coarse aggregate, relatively small-diameter cores may be taken, minimising the impact of sampling on the structure; cores as small as 50 mm diameter may be suitable. With concrete or concrete repair materials containing coarse aggregate 70 mm or larger diameter cores may be more appropriate.

It is always desirable that samples of concrete repairs include some of the concrete substrate. In some cases core sampling may cause the repair material to separate from the concrete substrate. Where this occurs it may indicate a weakness in the bond between the repair and substrate and it is desirable that the sample location be checked to see if it is hollow-sounding before core sampling commences.

Sampling by core drilling can be conveniently used in conjunction with pull-off testing as an effective means of providing information on bond strength at the same time as providing a sample suitable for petrographic examination.

Lump samples

Concrete lump samples obtained by disc cutting or with a hammer and chisel are less than ideal as they are likely to suffer damage during sampling. In some cases concrete repairs may have detached from the substrate and broken samples may be conveniently collected. However these may be of limited value as they may not include the concrete substrate and may have deteriorated since they became detached from the substrate. Samples of this type might also have been damaged or contaminated with chlorides since they detached from the structure.

2.1.2 Petrographic examination and allied techniques

Petrographic examination is a long-established technique for the analysis of concrete samples. ASTM C457(ASTM, 2010) and the Applied Petrography Group Code of Practice for the petrographic examination of concrete (APG-SR2, 2010) give standard procedures for the petrographic examination of concrete samples. A report from the Concrete Society also covers the topic in detail (CSTR No. 71, 2010).

Fundamental to petrographic examination is the preparation of thin sections for examination in transmitted light with a petrological microscope. Thin sections are samples prepared as thin slices typically 20–30 μm thick that are mounted on to glass slides and covered with glass cover slips. A typical thin section would cover an area of about 50 × 75 mm. The detail of the method of preparation of thin sections is beyond the scope of this publication and is described elsewhere (APG-SR2, 2010).

Thin sections of concrete are normally prepared using samples that have been vacuum impregnated with low-viscosity epoxy resin containing fluorescent dye and this facilitates the examination of microcracking and porosity distribution (illustrated in Figure 2.1) and avoids the possibility of introducing cracking during the preparation of the thin sections. It is desirable that in making the thin sections the exposure of the samples to water be kept to a minimum and that the samples are not heated to temperatures of > 45°C.

The thin sections are examined with a high-quality polarising petrological photomicroscope (illustrated in Figure 2.2) and these are generally fitted with dedicated digital microscope cameras. The microscopes employed for petrographic examination have sufficient resolution to clearly distinguish individual cement grains and the maximum resolution would be of the order of 2 μm. Modern petrological microscopes are equipped with a fluorescence illumination system and a range of objective lenses with magnifications ranging from ×20 up to ×1000. Stereo binocular microscopes with much lower magnifications are also essential for the examination of broken surfaces and larger scale features.

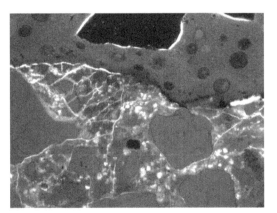

Figure 2.1 Thin section photomicrograph taken using fluorescent illuminations. It shows microcracks and areas of high porosity. The view shows a concrete surface at the top of the field of view that is covered by a layer of resin. Abundant microcracks are visible just below the broken surface of the concrete that result from the method of preparation of the concrete surface. The aggregate particles are of very low porosity and appear as dark grey particles in the lower part of the picture.

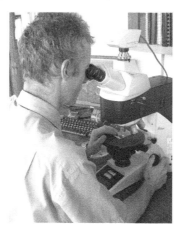

Figure 2.2 A modern high quality polarising petrological microscope. The Leica microscope shown is equipped with transmitted light illumination with infinity-corrected optics, and is capable of resolving objects as small as 2 μm.

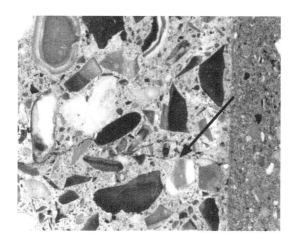

Figure 2.3 A polished surface prepared through a sprayed concrete repair. The sprayed concrete is visible on the right side of the field of view and there is a sharp contact with the concrete. The concrete contains a shrinkage crack (arrowed) which is truncated at the contact with the sprayed concrete.

Polished plates are invaluable for showing the large-scale structure of deep concrete repairs and the nature of the contact with the substrate. A typical polished plate is illustrated in Figure 2.3. Examining polished surfaces treated with fluorescent dye is a useful technique for determining crack geometry, especially when used with a stereo binocular microscope.

Among the types of information that can be gained from petrographic examination are:

- Aggregate and binder contents
- Water/cement ratio
- Aggregate type
- Aggregate deterioration such as ASR
- Binder type
- Binder deterioration such as leaching or sulphate attack
- Porosity and microcrack distribution

Evidence for deterioration at the repair/concrete contact.

2.1.3 *Electron microscopy and X-ray microanalysis (SEM/EDXA)*

Electron microscopy and X-ray microanalysis techniques are capable of providing information about chemical composition and microstructure at magnifications that are much higher than those that can be achieved with the petrological microscope. Electron microscopy is often most successfully employed when used in conjunction with conventional petrographic examination. Some examples of the application of the techniques of electron microscopy and X-ray microanalysis in the examination of concrete repairs would include:

- Mapping the penetration of chloride and sulphate ions
- Measuring PFA or GGBS contents
- Identifying the products of chemical attack
- Assessing the effectiveness of cathodic protection systems (Eden, 2003a).

The scanning electron microscope (SEM) uses a scanning beam of electrons to form an image of a sample. The electron microscope has the advantage that it is capable of operating at very high magnifications and can resolve objects that are much less than 1 μm in diameter. X-rays emitted as the electron beam interacts with the surface of the sample can be collected using an X-ray microanalysis system (EDXA) attached to the SEM and can be used to make quantitative chemical analyses of objects as small as 3 μm. The X-ray microanalysis system can also be used to map the distribution of elements around cracks or at interfaces with substrates.

2.2 Identifying the causes of concrete deterioration

Concrete repairs are commonly carried out in response to concrete deterioration. When examining concrete repairs that have been in place for several years, the original reason for the repair may not be obvious from a visual inspection of the structure, yet establishing the cause of the concrete deterioration is vital if reliable judgements are to be made about the potential life-span of a repair and the selection of appropriate repair materials.

Petrography is an effective means by which the causes of concrete deterioration can be established. Reinforcement corrosion is probably the most common reason for concrete repair; however, there are many other possible causes for concrete deterioration. The following sections list some of the more common causes of concrete deterioration (see also Chapter 1).

Chloride-induced reinforcement corrosion

In most cases the chlorides will be externally derived from sources such as road de-icing salts but it is essential to explore the possibility that the chlorides may have been originally incorporated in the concrete, as in some older concrete structures where calcium chloride may have been used as an accelerator. It is also important to bear in mind that carbonation may modify the distribution of chlorides in the concrete and lead to apparently low levels of chloride at the surface whilst chloride levels at greater depths may be sufficiently high to cause corrosion. Petrography is useful in establishing the pathways for chloride ingress and SEM/EDXA can be used to map out chloride distribution around cracks, analyse corrosion products and investigate the relationship between chlorides and carbonation. The corrosion products resulting from chloride-induced corrosion can be very fluid and substantial loss of reinforcement cross-sectional area can result from corrosion with little or no associated concrete expansion. Figure 2.4 shows the leakage of fluid reinforcement corrosion products along microcracks in the vicinity of corroded reinforcement affected by chloride ingress.

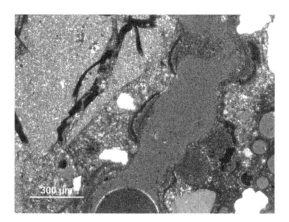

Figure 2.4 Thin section photomicrograph showing the contact between a patch repair mortar and a mechanically roughened concrete surface. The repair mortar has separated from the concrete surface and a crack between the repair material and concrete surface occupies the central part of the field of view. The crack surface is coated with reinforcement corrosion products and some of the cracks within the surface of the concrete substrate also contain corrosion products indicating on-going corrosion since the repair of the concrete surface. Note the shattering in the flint coarse aggregate particle visible on the left side of the field of view. There is also some patchy carbonation of the cement paste either side of the crack along the repair/ substrate interface.

Carbonation-induced reinforcement corrosion

Conventional testing with phenolphthalein will be sufficient in most cases to measure carbonation depth in Portland cement concretes. However, with HAC concrete and in some Portland cement concretes such as those containing SRPC, petrographic examination is the definitive means for measuring carbonation depth. It is important to establish whether or not there is evidence for moisture penetration since corrosion will not proceed in dry conditions. Carbonation and moisture-induced corrosion is generally associated with obvious expansion and spalling – particularly where reinforcement is situated close to concrete surfaces or intersected by cracks initiated by other causes such as drying shrinkage or AAR.

Alkali–aggregate reaction (AAR)

It is important to establish the aggregate type and whether or not AAR is associated with deleterious expansion. Figure 2.5 illustrates the typical appearance of concrete affected by AAR in a thin section.

SEM/EDXA can provide valuable information about the chemical composition of gel deposits that can be of assistance in assessing whether or not AAR is likely to be on-going. Core expansion testing can be used to measure the potential for future expansion. It is important that repair materials selected do not act as a source of alkalis and that the repair material contains aggregates that are of low reactivity (Concrete Society, 1999) with respect to AAR. Core sampling of concrete structures for the possible presence of AAR should sample the concrete at depth and should be at least 150 mm deep. The cores should be washed clean and immediately wrapped in several layers of cling film before dispatch to the laboratory. Petrographic examination is the definitive method for detecting AAR.

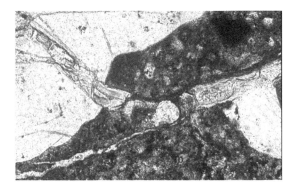

Figure 2.5 Thin section of concrete suffering from AAR, showing cracks infilled with alkali–silica gel.

Sulphate attack

Sulphate attack is most commonly encountered in environments where concrete is buried in ground containing sulphates. There are several forms of sulphate attack and establishing the type of sulphate attack is of importance in assessing the most appropriate repair materials. The thaumasite form of sulphate attack (TSA) has been documented in the case of some motorway structures on Liassic clay in Gloucestershire (Eden, 2003b) and equally

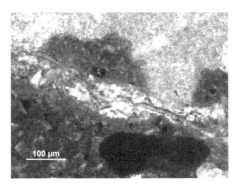

Figure 2.6 Thin section photomicrograph illustrating the thaumasite form of sulphate attack within a concrete repair material at the contact with a concrete substrate. The thaumasite occurs either side of a horizontal crack running from left to right and appears bright in the image. A limestone coarse aggregate particle occurs at the top of the field of view and the bottom of the field of view is occupied by the Portland cement based matrix in the repair material. The development of sulphate attack within the repair in this sample is related to the ingress of moisture containing sulphate ions along the interface between the repair and substrate.

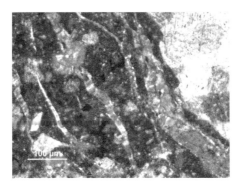

Figure 2.7 Thin section view showing the development of sulphate attack within a repair mortar. Note the abundant cracks infilled with ettringite crystals that appear mottled white to grey. The cracks occupy much of the field of view and are surrounded by porous paste that contains limestone dust. A limestone aggregate particle occurs on the right side of the field of view. The sulphate attack is related to the continued ingress of moisture containing sulphate ions along the repair/substrate interface.

affects concretes with or without sulphate-resisting Portland cement (SRPC). Sulphate attack is also common in concrete exposed to bacterially derived sulphur dioxide in sewage treatment structures. Figures 2.6 and 2.7 show the development of TSA and ettringite-related sulphate attack in a concrete repair. In many cases of sulphate attack the transition from severely weakened concrete to concrete showing little or no evidence for weakening is sharp and it is important that all the affected concrete is removed prior to instigating repairs. Petrographic examination can be used to measure the maximum depth of affected concrete.

Delayed ettringite formation (DEF)

This is a relatively rare cause of on-going damaging expansion encountered in concrete cured at elevated temperature – usually $> 65–75°C$ resulting from expansive ettringite formation in the cement paste in the presence of moisture. Controlling rates of moisture ingress may slow the rate of future concrete expansion and core expansion testing (BCA, 1988) can be used as a guide to the maximum future potential for expansion. Petrographic examination is one of the most reliable means of detecting this form of deterioration. With concrete placed *in situ DEF may only be present in the concrete at depths where temperatures reached during curing would have been at their greatest.*

Fire damage

Fire damage commonly results in spalling and the development of surface-parallel cracking. Fire damage may also modify the porosity of the cement hydrates and generate intense microcracking that may be invisible to the naked eye. In most cases it is the concrete that has been heated to temperatures $> 300°C$ that shows the most severe weakening. However, cracking may develop at much lower temperatures and at greater depths where reinforcement is affected by temperature increase and undergoes thermal expansion. A series of changes in the microstructure of the cement paste and aggregate particles can develop in concrete exposed to fire and it is possible by petrographic analysis to establish the depth of various temperature isotherms. The primary use of petrographic examination in examining fire-damaged concrete is in detecting the maximum depth of weakened concrete and the depth of concrete that needs to be removed before repairs can begin (Figures 2.8 and 2.9).

Freeze–thaw damage

Freeze–thaw damage occurs when water-saturated concrete is subjected to cycles of freezing and thawing and results from the expansion that occurs when water freezes within cracks or voids. In common with fire damage and sulphate attack, freeze–thaw damage tends to produce surface-parallel cracks.

In the UK the effects of freeze–thaw damage are usually limited to several centimetres depth, but in some places the effects of freeze–thaw damage may penetrate to several tens of centimetres. Petrographic examination can be used to measure the depth of deteriorated concrete and establish the depth of concrete to be removed prior to repair. The repair material needs to be resistant to moisture penetration and it is particularly important that the repair/substrate contact is sound with no potential for moisture build up. In some parts of the UK, aggregates susceptible to drying shrinkage may exacerbate the effects of freeze–thaw damage.

Frost damage strictly refers to damage to concrete when it freezes during the setting process and as such should not be confused with freeze–thaw

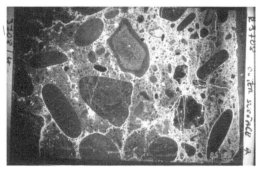

Figure 2.8 Fluorescent light photomicrograph of a thin section taken at low magnification to illustrate the development of microcracking in a concrete sample affected by fire damage. The external surface is visible along the right side of the field of view. Note the development of surface-parallel microcracks that pass through both the coarse aggregate particles and the surrounding paste. The paste towards the inner end of the sample has a patchy, moderate to low porosity.

Figure 2.9 Photomicrograph of a thin section taken at a depth of about 2 mm below the surface of fire-damaged concrete. The paste has developed intense microcracking and is substantially weakened. The level of damage seen in the section of this part of the sample is consistent with temperatures in excess of 300°C.

damage, which occurs after hardening. The two terms, unfortunately, are often used interchangeably.

Impact/mechanical damage and damage due to structural loading

These types of damage may be hard to distinguish on the basis of the examination of laboratory samples alone and site observations in conjunction with a structural assessment are required to unambiguously diagnose these types of damage.

2.3 Characteristics of some common cementitious repair materials

2.3.1 Hand-applied high-performance proprietary repair mortar mixes

Modern proprietary high-performance repair mortars can be used to produce very robust and durable repairs that benefit from low shrinkage characteristics and very low permeabilities. They commonly have complex compositions and can contain a very wide range of constituents in addition to Portland cement and aggregates of various types. Among the range of constituents that can be detected petrographically are:

- High-alumina cement/calcium aluminate cement
- Shrinkage-compensating cements
- Microsilica (densified, undispersed clots)
- PFA (pulverised fly ash) and PFA cenospheres
- GGBS (ground, granulated blast furnace slag)

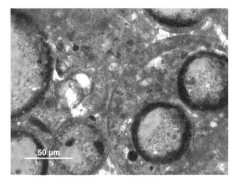

Figure 2.10 Photomicrograph of a thin section showing a proprietary repair mortar containing PFA cenospheres and particles of unhydrated Portland cement. The PFA cenospheres are hollow, spherical structures composed of glass and are abundant throughout the field of view. The unhydrated cement grain occurs just to the left of the field of view. There are also moderate quantities of portlandite crystals that appear pale to white.

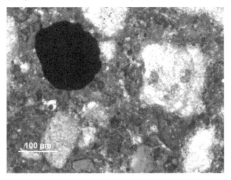

Figure 2.11 Photomicrograph view showing a sprayed concrete repair material containing graphite particles. A graphite particle occurs towards the top of the field of view and appears black. The repair material also contains crushed limestone filler and fine aggregate particles and occasional particles of GGBS visible towards the bottom of the field of view. The binder is based on Portland cement.

- Fibre reinforcement
- Graphite
- Inert fillers such as limestone dust.

In addition to the materials listed above, mortars of this type are also likely to contain polymers such as acrylic copolymers or Styrene Butadiene Rubber (SBR) and water-reducing additives that can only be reliably detected using methods such as infra-red spectroscopy. Examples of the proprietary repair mortars are shown in Figures 2.10 and 2.11.

2.3.2 Sprayed concretes (see also Chapter 13)

Sprayed concretes share a similar degree of complexity in composition to non-spray applied repair mortars but can often be readily distinguished from hand-applied mortars by a visual inspection of core samples on the basis of their characteristic internal layering (illustrated in Figure 2.12).

The internal layering commonly results from small variations in water and void content, but in some cases there may also be variation in aggregate distribution. The cause of the layering will be affected to a large extent by the method of mixing and application of the sprayed concrete. Sprayed concretes with water added at the nozzle can potentially show much greater variation in water content compared with many other types of sprayed concrete. Sprayed concrete repairs containing reinforcement may have concentrations of voidage behind the reinforcement caused by 'shadowing' during the spray application of the repair.

It is not uncommon for there to be a narrow binder-rich zone in sprayed concrete at the contact with the substrate that is caused by aggregate-rebound. Using thin sections it is possible to determine the causes of layering

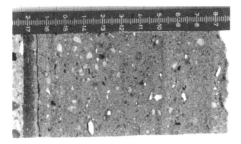

Figure 2.12 Polished surface view showing a sprayed concrete repair and the contact with the concrete substrate which occurs on the left of the field of view. Note the layered structure of the sprayed concrete repair.

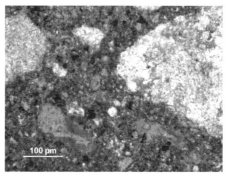

Figure 2.13 Photomicrograph showing the typical appearance of a sprayed concrete repair in thin section. The matrix is of low porosity and contains abundant particles of unhydrated Portland cement and small amounts of limestone dust. It also contains voids that are lined with small amounts of polymer. Limestone aggregate particles are visible on the left and right sides of the field of view.

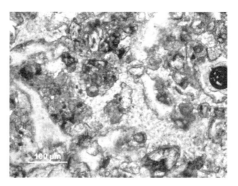

Figure 2.14 Photomicrograph of the same sample as Figure 2.13, but showing a layer of sprayed concrete with a locally high void content. The voids have taken in the resin used during the preparation of the thin section. The field of view is occupied by abundant particles of unhydrated Portland cement and areas of separated polymer that appear pale grey.

and the likely effects of the layering on the permeability and durability of the sprayed concrete repair. Figures 2.13 and 2.14 illustrate the typical appearance of sprayed concretes in thin sections.

2.3.3 Non-proprietary hand-applied mortars based on Portland cement

Repair materials of this type tend to be applied on a small scale. They range from dense, low-porosity mortars that might be expected to be as durable as carefully mixed and applied proprietary mortars through to very low durability mortars that carbonate rapidly and have high water/cement and aggregate/cement ratios. An example of a repair mortar of this type is shown in Figure 2.15.

2.4 Investigating the condition of the bond between a concrete repair and the concrete substrate

2.4.1 The effects of differing methods of surface preparation on bond strength and repair durability

Many failures of concrete repairs result from weaknesses of the bond between the repair material and the substrate. Petrographic examination can provide much information about the nature of the concrete surface and the method of preparation of the concrete surface.

Most repairs have a roughened substrate surface that is intended to provide a key for the bond with the repair material. The roughening of the concrete substrate is most commonly carried out either by mechanical means using a method such as scabbling or by water-jetting.

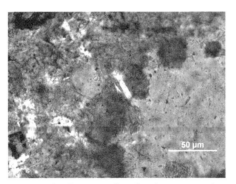

Figure 2.15 Photomicrograph showing the development of crystals of secondary ettringite within a void in a proprietary repair mortar. The development of such crystals is indicative of moisture ingress and recrystallisation of the cement hydrates. The ettringite crystals are needle-like and occur on the right side of the field of view within a void that contains the resin used during the preparation of the thin section. The left side of the field of view is occupied by moderate-porosity paste containing bright portlandite crystals and small quantities of unhydrated Portland cement.

Roughening of the surface of the structure to be repaired by mechanical means can leave behind a shattered zone in the substrate surface with intense microcracking invisible to the naked eye. Cracking of this type is readily detectable petrographically (illustrated in Figure 2.4) and can lead to a weakness in the bond between the repair and the substrate. Such microcracking can also act as a pathway for subsequent moisture ingress and weakening due to moisture movement and leaching.

Water-jetting is much less commonly associated with shattering of the concrete surface and commonly leaves a rough, cleanly broken surface.

The edges of concrete repairs may also have sawn margins or be feather-edged. Feather-edged repairs can show a tendency to lift at their periphery – especially if the concrete surface is smooth, allowing moisture and carbonation penetration. Figure 2.15 illustrates a feather-edged repair.

2.4.2 Common causes of post-repair weakening of the bond between the repair and the substrate

Moisture ingress

It is not uncommon for there to be a concentration of voidage or microcracking along the substrate surface. In damp conditions, these features can act as a pathway for moisture ingress leading to localised recrystallisation of the cement hydrates and porosity enhancement due to leaching of water-soluble materials in the cement hydrates such as portlandite. Figure 2.16 illustrates secondary ettringite formation in voids at the concrete surface due to recrystallisation of the cement hydrates. The moisture may also contain chloride ions contributing to the risk of the development of reinforcement corrosion.

Carbonation in damp conditions can result from the presence of carbonate ions in moisture that is able to penetrate into the repair. As well as

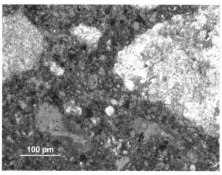

Figure 2.16 Photomicrograph of a basic repair mortar containing Portland cement and fine sand. The hydrated cement paste is of high porosity and is deeply carbonated and contains small quantities of unhydrated cement grains. The quartz aggregate particles appear various shades of grey to white.

contributing to the risk of the development of reinforcement corrosion, such carbonation can also have a weakening effect that is related to the unusually coarse texture of the calcium carbonate crystals formed in the 'carbonated' cement hydrates.

Stress concentrations due to differences in the strength and thermal expansion characteristics of the repair and substrate concretes

A very large range of proprietary repair materials are currently available, ranging from very dense, low porosity materials with a high compressive strength and modulus of elasticity through to relatively low strength and low modulus repair materials such as those containing PFA cenospheres and with high entrained air contents.

It is desirable for modulus of elasticity, compressive strength and thermal expansion characteristics of repair materials to be compatible with those of the substrate. For example, a dense, high modulus repair mortar applied to a relative weak substrate with a low modulus of elasticity may develop microcracking at the repair/substrate interface when the structure is subjected to structural loading or large temperature variations due to stress concentrations at the repair/substrate interface. With existing repairs, identifying the composition, porosity distribution and binder types in the concrete substrate and repair materials is an essential step to recognising compatibility problems of this type.

2.5 Common causes of cracking in concrete repair materials

2.5.1 Shrinkage

Shrinkage cracking falls in to two main categories:

- Drying shrinkage is a long-term phenomenon that will begin as soon as the concrete has hardened. The rate of drying shrinkage will decrease exponentially with time and a high proportion of the total possible drying shrinkage will occur within the first year after the application of the repair.
- Plastic shrinkage develops before the concrete hardens and whilst the concrete is still plastic.

Fine cracking and microcracking caused by drying shrinkage is one of the most common forms of cracking seen in the surfaces of concrete repairs. Cracking of this type is commonly most visible when the surface of the repair begins to dry out after being wetted. The potential for drying shrinkage is governed to a large extent by the cement and water contents of the repair materials and the high cement contents of many repair materials may contribute to the potential for shrinkage. Shrinkage-compensating

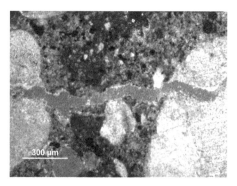

Figure 2.17 Photomicrograph showing a shrinkage fine crack developed in a sprayed concrete repair. The crack intersects the surface of the reinforcement and there is patchy carbonation either side of the crack shown by the very thin bright areas adjacent to the crack. Note the crack passes through the limestone aggregate particles as well as through the paste. Limestone aggregate particles occur on the left and right sides of the field of view.

cements will help control cracking of this type but may not entirely eliminate shrinkage cracking.

Drying shrinkage cracking is normally very shallow and the cracks may only be a few micrometres wide. However, where reinforcement is present within the repair it is not uncommon for shrinkage cracks to continue for several centimetres and to intersect the surface of the reinforcement. Shrinkage microcracks that are very shallow may self-anneal with time in damp conditions. Shrinkage cracking that intersects the surface of reinforcement may be detrimental to the long-term durability of the repair if they allow chlorides, moisture or carbonation to reach the surface of the reinforcement. Such cracking is readily detected using petrographic techniques; the example in Figure 2.17 shows the localised penetration of carbonation along a vertical shrinkage microcrack which intersects the surface of the, as yet, uncorroded reinforcement.

Plastic shrinkage can develop in the surfaces of concrete repairs where the repair is exposed to warm windy conditions before it has cured. It occurs where the rate of evaporation exceeds the rate of bleeding. Very thin repairs may be prone to cracking of this type, particularly where there is potential for moisture from the repair material to be absorbed by the substrate. Plastic shrinkage cracks are generally restricted to the cement paste, tend to be non-parallel sided and commonly decrease in width rapidly with depth.

2.5.2 The effects of on-going deterioration in the concrete substrate

In some cases expansion of the concrete substrate due to on-going deterioration can lead to cracking along the interface between the repair and the substrate – especially where the bond is already weak. In other cases cracks may be traceable as originating within the concrete substrate and continuing into the concrete repair. Figure 2.4 illustrates an example of ongoing deterioration in the concrete substrate due to the reinforcement corrosion affecting a concrete repair.

Failure to correctly diagnose or to take account of on-going deterioration in the concrete substrate will in many cases greatly reduce the lifetime of a concrete repair. The petrographic examination of core samples that include concrete substrate as well as the repair can be used to detect features such as cracking that originate within the concrete substrate and continue into the concrete repair.

References

Allman, M. and Lawrence, D.F. *Geological Laboratory Techniques*. London, Blandford Press, 1972.

APG-SR2, 2010 Applied Petrography Group, 'A code of practice for the petrographic examination of concrete.

ASTM C457/C457M – 10 'Standard Test Method for Microscopical Determination of Parameters of the Air-Void System in Hardened Concrete'. American Society for Testing Materials, USA, 2010.

ASTM C856 'Standard practice for the petrographic examination of hardened concrete'. American Society for Testing Materials, USA, 2004.

British Cement Association. *The Diagnosis of Alkali-Silica Reaction. Report of a working party*, BCA, 1988.

Concrete Society, Technical Report No 30, 'Alkali–Silica Reaction – Minimizing the risk of damage to concrete.' 3rd Edition, The Concrete Society, UK, 1999.

Concrete Society, Technical Report No 71, 'Concrete Petrography.' The Concrete Society, UK, April 2010.

Eden, M.A. 'Investigation Into the Effectiveness of the CP System in the Replaced RC Road Deck: West Tunnel, Dartford River Crossing'. *Proceedings of Concrete Solutions, 1st International Conference on Concrete Repair*, St. Malo, France. GR Technologie, Barnet, UK, 2003a.

Eden, M.A. 'The laboratory investigation of concrete affected by TSA in the UK', *Cement and Concrete Composites*, Vol 25 (2003b) 847–850.

Eden, M.A., White, P.S. and Wimpenny, D.E. 'A petrographic investigation of concrete with suspected delayed ettringite formation – a case study from a bridge in Malaysia'. 11th Euroseminar on Microscopy Applied to Building Materials, 5–9 June 2007, Porto, Portugal.

French, W.J. 'Concrete petrography: a review'. *Quarterly Journal of Engineering Geology* Vol. 24 (1), (1991) 17–48.

St John, D.A., Poole, A.B. and Sims, I. *Concrete Petrography, A Handbook of Investigative Techniques*, Edward Arnold, London, 1998.

3 Structural aspects of repair

Jonathan G. M. Wood

3.1 Structural repair of concrete

Reinforced and prestressed concrete are structural materials. Two major failures (Wood, undated, accessed July 2010; Wood, 2003, 2008) following inadequate repair contracts have highlighted the importance of structural consideration of both deterioration and repair. When damage or deterioration has occurred necessitating a repair, the changes to the structure before, during and after the repair need to be a primary consideration. This will throw up a range of questions about the structure and how the actual build quality relates to the designers' intentions, current standards and how deterioration has altered this.

The IStructE publication 'Appraisal of existing structures' (IStructE, 2010) deals comprehensively with the process of structural appraisal. If

Figure 3.1 De la Concorde overpass after collapse, five dead. The deteriorated badly repaired half joint and lower part of the cantilever can be seen hanging down.

reinforcement drawings are available, they will need to be validated with checks on reinforcement size, location and cover. If they are not available, it is essential that they are recreated from surveys. This is very expensive and underlines the value to owners of maintaining full records of construction, modifications and repairs.

The structural reinforcement cover surveys are complementary to and should be integrated with the investigations to evaluate cover in considering corrosion risk. However, missing bars and top reinforcement cover greater than specified will go unremarked in a corrosion survey, but they may indicate serious structural deficiencies even before deterioration is considered.

When damage or deterioration becomes obvious to the owner, the first questions relate to overall safety and the usually more immediate risk to the public from falling spalls. So, structural engineering input (Wood, 2006) to the investigation and subsequent repairs is essential. The amount of deterioration concrete structures can suffer before remedial measures are necessary is small. Codes for concrete design include no margin for deterioration and modern design to minimise costs has reduced strength reserves and robustness.

Focusing the investigation on the areas where there is structural risk from vulnerable details and/or the worst deterioration usually enables a better and more cost-effective diagnosis to be made than random sampling. Figure 3.2 shows inadequate bearing seating of precast beams below a leaking car park joint.

Careful evaluation of the causes of cracking is an essential procedure in the appraisal of a concrete structure. Cracking arises from:

- Structural tensile strains from flexure and shear
- Early age 'non-structural' effects, thermal, shrinkage, etc
- Long-term 'non-structural' conditions corrosion (Concrete Society, 1992), AAR (BRE, 2007, IStructE, 1992, 2010), etc.

Relating crack widths to causes and monitoring changes in crack widths are powerful diagnostic tools. Any area where tensile strains exceed about 150 microstrain will crack and the crack orientation indicates the principal stress direction. Compressive stresses suppress cracking. Cracking relates to the combination of all structural and 'non-structural' effects. For example, the axial compression in a column suppresses map cracking from AAR so only vertical cracking develops. Cracking provides a channel for the ingress of chlorides and carbonation.

Sets of cores taken during an investigation which contain cut reinforcement immediately raise alarm bells for the structural engineer. Not only is it extremely difficult to structurally reconnect a cut reinforcing bar, but the structural damage from cutting a bar may be far greater than that from the deterioration being investigated. All coring proposals need to be checked by a structural engineer and coring should be preceded by a cover survey of

Figure 3.2 Inadequate seating of precast beams below a leaking car park joint (Wood, 2009).

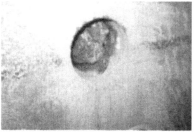

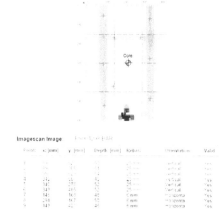

Figure 3.3 Ferroscan and 50 mm core sampling with pull-off tests of repairs on Tuckton Bridge (from Wood, 2008).

the area to ensure that the core goes between reinforcing bars, not through one. The core size should be minimised to that which is essential for tests, typically 50 mm or 70 mm.

Once the reinforcement layout and visual signs of corrosion at spall locations are apparent, a sensible estimate, guided by selective opening up, needs to be made of the extent of corroding bars which are not yet spalling the cover. Corrosion of a bar near the surface will cause a spall, but if the bar depth is more than twice its diameter there will be a very severe loss of section before a spall can develop. With closely spaced deep bars delamination of an area can develop, rather than individual spalls above bars.

One needs to consider the likely condition of steel in hidden concrete behind cladding and in joints where carbonation and/or chloride ingress will have developed. Periodic moisture from condensation, driving rain and/ or due to failing sealants can then accelerate corrosion. Some cladding has hidden fixings of mild steel, which will corrode, or galvanised steel which last only a little longer or of stainless steel which is good, as long as there is no bimetallic corrosion.

When concrete is saturated, which limits oxygen availability, and there is a high chloride concentration, corrosion develops as 'black rust' without the expansion to produce spalling. This makes it difficult to detect. Structurally 'black rust' is important as it tends to cut bars locally, (Figure 3.4), especially the bends on the shear links and column stirrups which rapidly reduces the beam or column strength.

Figure 3.4 Local pitting corrosion of links in a half joint, from chlorides in saturated concrete.

Half-cell surveys help identify the areas of active corrosion on that day and corrosion rate measurements give an instant number, but they give no indication of the severity of reinforcement section loss. Both work best when concrete is damp. They can help locate areas for selective opening up to answer the structural question: 'How much loss of section and bond strength at laps has occurred over the life of the structure?'

Selective opening up can give a better indication of current damage and the likely extent of extra cutting out for repairs to effectively control corrosion. Where conditions are right cathodic protection (CP) can reduce the need for cutting out. However before applying CP the safety of the already corroded structure must be checked.

It is an unusual repair contract which does not reveal that deterioration is more extensive than was apparent from the initial investigation. There is a strong case for having a trial repair contract to enable more extensive and deeper destructive investigation and evaluation of the effectiveness of repairs before a large-scale contract is let.

Effective repair requires cutting out to a sound substrate and to clear all carbonated and chloride contaminated material around the reinforcement (Figure 3.5). This cutting out inevitably further weakens the structure and the effects of this must be evaluated before the contract is let, including the potential consequences of more extensive cutting out, if that is found to be necessary as work progresses.

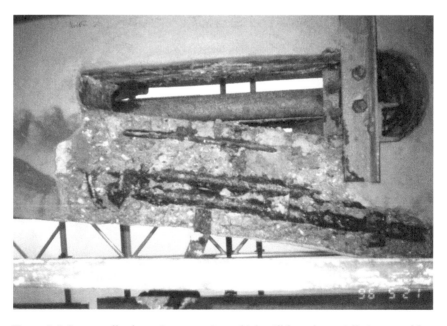

Figure 3.5 Structurally damaging corrosion which will be substantially increased by cutting out to clear chlorides.

The cost of access and disruption associated with repairs makes it important to be aware of the extent to which further deterioration may develop over the next 20 years and its structural consequences (Wood, 2009). Tuckton Bridge, built in 1905, is a good example of progressive cycles of repair (Wood et al., 2008). Often deterioration has been aggravated by bad local details creating ponding or channelling moisture and salt into joints. Upgraded drainage and providing shelter to slow deterioration should be considered in the prediction of future deterioration.

Extrapolation from the statistical analysis of cover depth related to chloride and/or carbonation profiles, including the data from the deeper investigation in the trial repair contract, should enable future trends to be predicted. The likely cost of these future repairs, as well as the currently proposed repairs, relative to the value of the structure must be considered before embarking on extensive remedial work. The poor performance of many concrete repairs, with 50% failures reported within 10 years (Matthews and Morlidge, 2006), needs to be considered in the prediction.

Some concrete structures are highly stressed with thin members with little margin for deterioration. Corrosion in exposed columns creates particular difficulties as this can involve cutting out behind all stirrups and main bars leaving just a central core of uncontained concrete, which makes removal and full recasting a more reliable option than patching. With precast construction, the joints may be badly constructed and poorly toleranced and they are particularly vulnerable to deterioration.

Other forms of concrete structure, sometimes for architectural effect, use large low-stressed members which can be cut into with little detrimental effect. An evaluation of stress levels and reinforcement configuration in a member will give a preliminary indication of the need for achieving a structural repair. For cosmetic reasons and/or corrosion control the repair material does not necessarily need to reinstate the load-carrying role of the original concrete, but it is important that it stays safely attached.

Where the reinforcement is corroded the structural role of the bars needs to be reconsidered relative to their current condition. Appraisal may have shown inadequacies in the original design relative to current requirements for shear strength, robustness and/or impact resistance. It may be appropriate, and in other cases essential, to remedy these shortcoming as part of the overall remedial works.

Large concrete members often contain reinforcement for early age crack control. If these bars with low covers are corroding, it may be appropriate, subject to a structural check, to cut some of them out to facilitate repair.

The three golden rules for design also apply to concrete repair. One must balance the three key objectives of:

• Fitness for purpose, including restoration of strength and control of further deterioration

- Cost effectiveness, including minimising disruption costs to the users of the structure
- Delighting the eye, either restoring to match the original concrete or providing an overall configuration and finishes to suit the continuing use.

Cutting out and recasting concrete to be structurally effective is difficult, time consuming and expensive. Sometimes it is simpler to provide a 'belt and braces' structural bypass, minimising the cutting out and leaving the original material to continue to deteriorate. In other circumstances, the complete removal of the element and recasting of the whole will be simpler, cheaper and structurally more effective then patching. There can be no standard rules: the remedial strategy and the detailing must be developed on a case-by-case basis.

3.2 Why is a full structural concrete repair difficult?

It has been reported by CONREPNET (Tilly and Jacobs, 2007, see also Chapter 14) that about 50% of concrete repairs fail within 10 years, but the fundamental causes of these failures are not fully clarified. These 'repairs', at considerable expense, have only served to recreate the risk of spalling. Structural repairs that fail are doubly bad as the cutting out to repair will have reduced strength, so that when they fail the structure is more seriously weakened.

When specific repair failures are analysed and related to the properties of available materials, it becomes clear that a reliable full structural repair requires:

- Cutting out to a sound substrate
- Removal of carbonated or chloride-contaminated concrete from around reinforcement, unless CP is viable
- Incorporating additional reinforcement or replacing it internally and/ or externally
- Supporting the structure while the repair is placed and gains its strength and stiffness, so that the repair carries a sufficient share of dead and live load
- Matching the physical and chemical properties of the repair mix to the original concrete
- Placing the repair mix to fully fill the cavity and key and bond to back and edges
- Finishing it in a way appropriate to the use of the structure.

3.2.1 Cutting out to a sound substrate

Analysis of repair failures shows that bruising of the substrate concrete is often a cause of low strength and delamination. A low pull-off test value only gives an indication of a problem, not a diagnosis. Ensuring that the coring for the test goes one diameter beyond the repair/concrete interface and removing this deep section, enables both sides of the pull-off fracture to be examined and a petrographic examination of a longitudinal section to be made. The tensile strength of reasonable concrete is typically 3 to 4 MPa, but variable in the pull-off test. When concrete is bruised by a jack hammer there is a weak surface of cracked material, (Figure 3.6), and this is where the pull-off failure often occurs. For structural repair, high-pressure water jetting can give an excellent surface. Where the underlying concrete is inherently weak from a poor mix, a stronger repair will simply pull off and this limits the structural effectiveness.

3.2.2 Removal of carbonated or chloride-contaminated concrete

The removal of carbonated or chloride-contaminated concrete needs to be considered in depth and for the surrounding area. An inherent characteristic of concrete and of the conditions of exposure, which strongly influence chloride ingress as well as carbonation, is their high variability. Combined with this is the variability of cover depth. It is important to take sufficient samples to get a true distribution. Spot samples can be very misleading.

The durability surveys will have given an indication of the overall patterns of corrosion. This should differentiate localised problems with shallow bars and carbonation of 15 mm maximum from those areas where carbonation depths are similar to the average reinforcement bar depth (cover + half diameter). It will also have identified if chlorides are contributing to

Figure 3.6 Weakened concrete bruised by a jack hammer – where the failures occur.

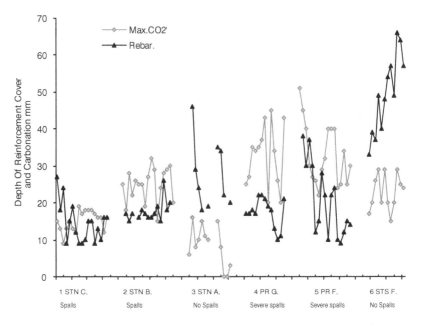

A3 - 2-4. Comparison of Max. Carbonation with Min.Cover to
Reinforcement for Six Detailed Survey Locations

Figure 3.7 Typical variation in carbonation depth and cover in one part of a structure.

the corrosion and in particular if calcium chloride is present beyond the carbonation front.

When cutting out for a specific repair the rebar depth and the extent of corrosion become clear. The carbonation depth can be easily checked with phenolphthalein, but site checks for chlorides are more difficult. The corrosion boundary on reinforcement is a guide to where current chloride conditions are on the threshold for corrosion, but beware incipient anodes. From this, an extrapolation for the next 20 years, with a margin for error, can be made.

The profiling of the cut-out area for repair also needs to consider the effectiveness with which the repair can be placed and its orientation with the stress fields during repair and when dead load and live load are reapplied.

3.2.3 *Incorporating additional reinforcement or replacing it*

The essence of a structure is its ability to transfer shear, flexure and axial compression from applied actions (dead and live load) to the ground and to provide a sufficient degree of ductility and robustness. Some reinforcement becomes redundant after construction or is oversized when appraised in detail. Some structures are structurally inadequate because of errors in old design codes, design errors and construction errors even before deterioration sets in.

The easiest way to understand structural behaviour is to consider the flow of forces through the structure from the applied forces to ground. Never forget longitudinal and lateral forces from wind and considerations of stability. Concrete and repairs can only reliably take compression. When a member, or one side of it, is in tension from axial load, bending or shear there must be adequate reinforcement to carry all that tension with particular attention to connections and laps. Corrosion will weaken the tensile load paths especially at connections.

Spalling and cutting out for repairs will reduce the compressive load paths. Only if dimensional compatibility of the repair with the underlying concrete is achieved and the reapplied load puts the repair into compression will it contribute to the strength of the structure (Emberson and Mays, 1990 & 1996). Repair patches which shrink and creep in the long term will be structurally ineffective and the strength will be that of the structure with the hole cut out for the 'repair'. The bigger the cut out, the weaker the structure becomes.

Some reinforcement is required by the reinforcement detailing rules, for example, stirrups in columns. They are not stressed directly but are essential to contain bursting and splitting of a column and to prevent buckling of the main reinforcing bars. Shear steel in beams ensures a ductile failure mode. When these links are corroded, the element becomes brittle with a risk of sudden failure. When reinforcement area is lost by corrosion, ductility and robustness must be considered as well as simple evaluation of stress.

Trying to fit new reinforcement into old concrete sections is never easy. Sufficient cover must be achieved. Stainless steel cannot be used in contact with normal reinforcement because of bimetallic corrosion. Lap connections need long lengths and the structure must be unloaded for new reinforcement to take its share of load.

Where deterioration is severe but localised, external strengthening (Figure 3.8) or the provision of an alternative load path around the weak part can be considered, but it is not easy to do this elegantly.

A column with corrosion of the stirrups may be better treated by wrapping with carbon fibre or steel bands to replace the containing function of the stirrups. A more seriously damaged column can be reinforced with non-structural patch repairs and steel plates bolted and epoxied on.

Since the 1980s the use of epoxy resins to attach stainless steel plates and carbon fibre for flexural strengthening of deficient reinforced and prestressed structures has become well established. However, when corrosion is damaging a structure, these techniques are inappropriate unless testing has established that there is no risk of corrosion delaminating the concrete cover to which they adhere. They can also restrict oxygen access and can aggravate 'black rust' if there is a sufficiently large cathodic area elsewhere to drive the corrosion cell.

Figure 3.8 Through bolting and clamping of splitting bridge tower head with AAR and deficient designed containment reinforcement.

3.2.4 Supporting the structure while the repair is placed

Simply cutting out and filling the hole with a repair material will often be ineffective for a structural repair, as the repair will not contribute to carrying dead load and shrinkage and creep often result in it carrying little live load. Supporting the structure during repair (Canisius et al., 2002) so that it fully gains strength and stiffness before load is reapplied will enable it to contribute more fully to overall strength.

Common prudence as well as Health and Safety require support to carry load and ensure stability when structures are being opened up for repair. This is particularly important when the degree of deterioration is uncertain and cutting out may extend beyond what was originally envisioned. Flat slabs around column heads of all types need particular caution.

3.2.5 Matching the repair to the concrete

The fundamental need for strain compatibility of repair (Figure 3.9) was spelt out nearly 2000 years ago, in the Bible, in Matthew 9:16.

> No man putteth a piece of new cloth unto an old garment, for that which is put in to fill up taketh from the garment, and the rent is made worse.

Figure 3.9 Differential shrinkage.

The principles of this are implicit in the overall approach of the BS EN 1504 Part 9 Eurocode on Concrete Repair (BSI, 2008). The overall approach evolved from work started in the 1980s by CIRIA (Davies, 2010, Pullar-Strecker 1987). However, the other parts of the standard have become tangled with a plethora of tests and CE (conforming to EU consumer safety, health or environmental requirements) ratings which do not facilitate the selection of effective structural repairs matched to a particular structure. Many of the properties of CE materials are geared for ease of application and fast strength gain to aid the contractor on site, but these tend to be detrimental to the long-term performance of the repair.

Matching repair properties with the structure concrete should be the primary requirement in any repair specification. Many proprietary CE repair products comply with EN 1504 but will be a poor match for most concretes. The EN needs to be used to assess the properties so that a match can be achieved.

Firstly, Young's modulus and the coefficient of thermal expansion need to match. A starting point for this is the use of cement and aggregates similar to the original. This will also help match the finished appearance.

It is tempting, but unhelpful, to produce a repair material which is markedly superior to the original material in terms of strength, carbonation resistance, permeability and diffusion resistance. Achieving better performance in the repair than the surrounding un-repaired areas will not extend the overall life of the structure. The more 'durable' material will have differential heating and wetting and drying characteristics, which can produce differential strains which will tend to delaminate the repair.

The dominant cause of delamination and failures of repairs to carry their share of loads on the structure is differential shrinkage as the repair dries into equilibrium with the adjacent concrete. Shrinkage depends on water cement ratio (w/c) and aggregate cement ratio (a/c) as shown in Table 3.1. So a repair mix needs to minimise water cement ratio and maximise aggregate cement ratio. Except for the largest repairs, there is advantage in reducing the maximum aggregate size to 10 mm to aid compaction. This, with an aggregate cement ratio of 6, should be the starting point for mix development. To keep the water cement ratio low and to aid compaction and adhesion, limited amounts of admixtures (super plasticiser or SBR) and adjustments to cementitious materials can be used subject to maintaining compatibility, with a cement content similar or a little less than the original.

Test data on long-term shrinkage is needed, as strain v. time plots for proprietary mixes are seldom supplied with product details. While water cement and aggregate cement ratios achieved in site repair trials are a guide to this, test data should always be obtained for the site mix. The CIRIA (1993) test is simple and robust and provides results which can be directly related to long-term site performance. When trial repairs are carried out sample prisms should be monitored for shrinkage on site where the drying conditions match those of the structure.

A perennial delusion is that expansive cements will compensate for long-term shrinkage. The expansion from the development of ettringite over the period of hydration, typically the first 2 days, needs to occur before the paste has hardened to the stage where the ettringite damages it. In a contained

Table 3.1 Shrinkage (microstrain) of prisms to 50% relative humidity (Neville, 1995, Table 9.3). These shrinkages need to be compared to the tensile strain to cracking for concrete and mortars of typically 150 to 200 microstrain.

At w/c *a/c*	*0.4*	*0.5*	*0.6*
3	800	1200	–
4	550	850	1050
5	400	600	750
6	300	400	550

situation such a mix will develop a residual compressive strain which will offset subsequent shrinkage, particularly where damp conditions reduce subsequent shrinkage. Where the hydrating mix is not well contained, as in most repairs, the expansion occurs outwards so little compressive strain is developed. In the dry conditions for many repairs, the long-term shrinkage far exceeds the short-term residual compressive strain. Late formation of the ettringite can damage the repair.

The inclusion of fibres in repair mixes can help keep the repair in one piece. However as failure occurs around the boundary of the repair, fibres give no advantage in preventing the edge and delamination cracking.

3.2.6 *Placing the repair mix*

The adhesion of the repair to the substrate is of great importance as this is where most repairs fail. Pre-moistening the substrate will prevent it sucking moisture out of the repair, but it is important to avoid over-wetting which gives a weakened repair interface. A bond coat compatible with the repair mix is beneficial, as long as it does not dry and harden before the repair is applied.

Placing the repair and keeping it in place as it bonds and gains strength requires tailored shuttering. The cohesion and consistency should ideally allow well-rammed filling, with care as it is applied, so that air is not trapped at the interface. Shuttering can than be applied to hold and compress the repair while it fully hydrates and gains strength. Early drying shrinkage and incomplete hydration can lead to early failure.

Well-controlled shotcrete can be an effective way of placing concrete to fully fill the cavity and key and bond to the back and edges. However, it is only economical where large areas of repair are needed.

3.2.7 *Finishing*

In some locations an approximate matching of profile and colour is sufficient. However where the concrete is an architectural feature and particularly where texture, board marking or exposed aggregates are a special feature, special measures are required. Often a two-layer approach is necessary, with the main repair providing structural/corrosion control and a finishing cosmetic layer matched to restore the original appearance.

3.3 Conclusion

Where deterioration or damage necessitates repair, there needs to be a structural engineering consideration in investigation, appraisal, specification and methodology of the repair. Cutting out for repair further weakens the structure and patching will not restore strength unless rigorous procedures are followed.

3.3.1 Check list

The following framework may be helpful as a check list for those involved in assessment, inspection and repair of deteriorating structures.

1   Check as-built drawings and maintenance records.
2   Carry out a structural review to identify the key areas of structural weakness and/or structural sensitivity to deterioration as a basis for inspection procedures. This should cover both strength and risks from spalling.
3   Check for any features:
   • for which factors of safety may be inadequate for actual construction method and quality
   • where the structural form is not explicitly covered by Codes which may be vulnerable to progressive collapse.
4   Establish by inspection and testing:
   • any departures from as-built drawings
   • any indications of defective or substandard construction
   • indications of severe local environments from ponding, waterproofing breakdown, seepage etc.
   • the current trends of deterioration and likely long-term trends.
5   Identify where and when protection, strengthening and/or repair may become appropriate and cost effective as part of the long-term maintenance programme.
6   Ensure that before repairs are carried out there is:
   • a full specification and procedure for repair, propping and testing
   • a Structural Engineer's check of the structure:
      • 'as built'
      • 'as deteriorated'
      • 'as cut out for repair, with propping if required'
      • 'as repaired'
      • 'with repair delaminated'.
   • The cost implications of maintenance and further repairs, estimated from deterioration trends over the next 20 years, have been considered.
7   Insist on a full recorded survey of condition before problems are hidden below patch repairs, coatings or waterproofing.

References

BRE Digest 'Corrosion of steel in concrete' (3-part set) (DG444) in BRE 'Concrete Pack' 2007.

BS EN 1504-9:2008 'European Standard on Concrete Repair. General principles for use of products and systems'. BSI, London, 2008.

Canisius, T.D.G. et al., *Concrete Repair Patches Under Propped and Un-Propped Conditions* FBE Report 3, BRE, March 2002.

CIRIA *Standard Tests for Repair Materials and Coatings for Concrete*, Part 1 'Pull-off tests' TN 139, Part 2 'Permeability tests' TN 140, Part 3 'Stability, substrate compatibility and shrinkage tests'. TN 141, 1993.

Concrete Society. Technical Report No 22, 'Non-Structural Cracking', The Concrete Society, 1992.

Davies, H. *Protection and Repair of Concrete – A Practical Guide to the European Standards*, London, Spon Press, 2010.

Emberson, N.K. and Mays, G.C. 'Significance of property mismatch in the patch repair of structural concrete', Parts 1 and 2. *Magazine of Concrete Research* Vol 42, No 152, pp. 147–170, September, 1990, and Part 3 *Magazine of Concrete Research* Vol 48, No 174, pp. 45–57, March 1996.

IStructE. *Structural Effects of Alkali-Silica Reaction – The Appraisal of Existing Structures.* London, SETO, 1992, with addenda 2010.

IStructE. *Appraisal of Existing Structures.* 2nd edn. 1996, 3rd edn. 2010.

Matthews, S.L. and Morlidge, J.R. 'What is wrong: Concrete repair – Solution or problem,' pp. 3–10 and associated CONREPNET papers, M. Grantham et al., Ed. *Proc. Concrete Solutions Conference St Malo*, BRE Press, 2006.

Neville, A. M. *Properties of Concrete*, London, Prentice Hall, 1995.

Pullar-Strecker, P. *Concrete Reinforcement Corrosion: ICE Design and Practice*, Thomas Telford 2002, updated from Pullar-Strecker, P., *Corrosion Damaged Concrete Assessment and Repair*, CIRIA, Butterworth, 1987.

Tilly, G.P. and Jacobs, J. *Concrete repairs – Performance In Service and Current Practice*, European Commission, London, BRE Press 2007.

Wood, J.G.M. 'Pipers Row Car Park Collapse: Identifying risk'. *Proceedings of an International Conference, Structural Faults and Repair*, Edinburgh, 2003. Summary in *Concrete Magazine*, Oct. 2003.

Wood, J.G.M. 'Evaluation of structural risk from concrete deterioration and repair', M. Grantham et al. eds, *Proc. Concrete Solutions Conference, St Malo*, 2006.

Wood, J.G.M. 'Implications of collapse of de la Concorde Overpass', *The Structural Engineer* Vol. 86, No.1, pp.16–18, 8 January 2008.

Wood, J.G.M. 'Improving guidance for engineering assessment and management of structures with AAR', *Proc 13th ICAAR*, M. Broekmans and B. Wigum ed., NTNU Trondheim, June 2008.

Wood, J.G.M. 'Structural repair of defects and deterioration to extend service life', in Grantham, M.G. ed. *Concrete Solutions, 2009*, 3rd International Conference on Concrete Repair, Padua, 2009.

Wood, J.G.M. 'Pipers Row Car Park, Wolverhampton: Quantitative Study of the Causes of the Partial Collapse on 20th March 1997', SS&D Contract Report to HSE, http://www.hse.gov.uk/research/misc/pipersrow.htm. Accessed July 2010.

Wood, J.G.M. Grantham, M.G. and Wait, S., 'Tuckton Bridge, Bournemouth. An Investigation of Condition after 102 years'. *Proc. Structural Faults*, 2008.

4 Cathodic protection of structures

John P. Broomfield

4.1 Impressed current cathodic protection

Cathodic protection is the process of controlling corrosion on a metal structure (such as a reinforcing cage or a pipeline) by turning it into a cathode. This is done by introducing an external anode into the conductive electrolyte that surrounds them both (such as concrete, soil or sea water). The terms anode and cathode are explained in the next section of this chapter.

There are two types of cathodic protection, one called impressed current cathodic protection, the other called galvanic or sacrificial anode cathodic protection. Galvanic cathodic protection was discovered in the 19th century by Sir Humphrey Davy and is very widely applied to structures in soils and water, especially sea water. It is closely related to the process of galvanising. Impressed current cathodic protection was invented far later but has been successfully used for a longer period on reinforced concrete structures. This is because of the high electrical resistance of concrete compared with soil or sea water requiring the higher voltage available to impressed current systems compared with galvanic anodes.

4.1.1 Corrosion theory

Concrete is a porous material. In atmospheric exposure conditions the pores will contain some water and some air which will be in contact with the embedded reinforcing steel. Steel in neutral or acid conditions will corrode in the presence of oxygen and water. However, concrete pore water contains sufficient alkalis (sodium, potassium and calcium hydroxides) to ensure that the oxide formed is a protective passive oxide layer which once formed, slows the corrosion rate to very low levels leading to excellent durability of reinforced concrete.

Concrete can therefore be considered to be a protective coating to the reinforcing steel. If that coating is damaged, by impact, excessive structural loads, erosion or chemical attack, the steel may be exposed and can corrode. However, there are two processes that will attack the reinforcing steel without first causing any damage to the concrete.

The first process is carbonation. It might be considered as concrete's equivalent to the oxidation of steel. Steel is formed by the removal of oxygen and other oxidising elements and compounds from iron ores. Concrete (or more accurately cement) is formed by driving off carbon dioxide from the constituents of cement clinker. Just as corrosion is the process of the oxygen recombining with steel or iron to revert back to the lowest chemical energy state, so the reaction of concrete with carbon dioxide is a comparable reaction. The difference is that iron oxide has drastically different (inferior) properties compared to steel while carbonated concrete has very similar properties to uncarbonated concrete except that it loses its alkalinity. This loss of alkalinity means that embedded steel becomes vulnerable to corrosion as the protective passive oxide layer breaks down.

The other process that will lead to reinforcement corrosion is chloride ingress. Chloride ions from sea salt, de-icing salt or from contaminated concrete mix constituents will break down the passive layer without damaging the concrete once there is a sufficient concentration at the steel surface along with sufficient oxygen and water.

Regardless of the cause of depassivation, the corrosion process proceeds in the same manner for carbonated and chloride-contaminated structures. Iron dissolves at the anode, forming Fe^{2+} irons and liberating electrons:

The anodic reaction: $Fe \rightarrow Fe^{2+} + 2e^-$ 　　　　　　　　　　　(4.1)

The two electrons ($2e^-$) created in the anodic reaction must be consumed elsewhere on the steel surface to preserve electrical neutrality. In other words we cannot have large amounts of electrical charge building up at one place on the steel. There must be another chemical reaction to consume the electrons. This is a reaction that uses water and oxygen to create alkaline hydroxyl ions:

The cathodic reaction: $2e^- + H_2O + \frac{1}{2}O_2 \rightarrow 2OH^-$ 　　　　　(4.2)

This is illustrated in Figure 4.1. The hydroxyl ions increase the local alkalinity and therefore will strengthen the passive layer, warding off the effects of carbonation and chloride ions at the cathode. Note that water and oxygen are needed at the cathode for corrosion to occur, but not at the anode. The ferrous ions go on to react with hydroxyl ions, water and oxygen, to create rust which has approximately seven times the volume of the original steel.

We can see from the reactions and the figure that there is a benign cathodic reaction producing hydroxyl ions, i.e. alkalinity. If we can force this to occur across the reinforcing steel network and displace the anodic reaction elsewhere, we will protect the steel from corrosion. An impressed current cathodic protection system does this by installing an anode system on the concrete surface or embedded in the concrete and applying a small

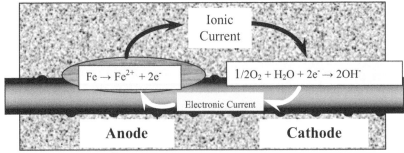

$$Fe^{2+} + 2OH^- \rightarrow Fe\,(OH)_2 \quad \text{Ferrous Hydroxide}$$

$$4Fe(OH)_2 + O_2 + 2H_2O \rightarrow 4Fe(OH)_3 \quad \text{Ferric Hydroxide}$$

$$2Fe(OH)_3 \rightarrow Fe_2O_3.H_2O + 2H_2O \quad \text{Hydrated Ferric Oxide (rust)}$$

Figure 4.1 The anodic, cathodic and iron hydration reactions for corroding steel.

electrical current: injecting electrons into the steel that forces the cathodic reaction to predominate.

There are two cathodic reactions that can occur on the steel surface. As well as hydroxyl production (equation 4.2) there is also a hydrogen evolution reaction which occurs if the electrochemical potential of the steel is too negative:

$$H_2O + e^- \rightarrow H + OH^- \tag{4.3}$$

This hydrogen evolution reaction can lead to hydrogen embrittlement and severe precautions must be taken when applying impressed current cathodic protection to prestressing or to structures that contain prestressing. Advice on such applications can be found in Broomfield (2007) and NACE (2001).

The chloride ion itself is negative and will be repelled by the negatively charged cathode (reinforcing steel). It will move towards the (new external) anode. With some anodes (particularly the carbon-based anodes) it may then combine to form chlorine gas at the anode:

$$2Cl^- \rightarrow Cl_{2(gas)} + 2\,e^- \tag{4.4}$$

The other major reaction at all major anodes is the formation of oxygen:

$$2OH^- \rightarrow H_2O + \tfrac{1}{2}O_2 + 2e^- \tag{4.5}$$

and

$$H_2O \rightarrow \tfrac{1}{2}O_2 + 2H^+ + 2e^- \tag{4.6}$$

Reaction (4.5) is the reverse of reaction (4.2), i.e. alkalinity is formed at the steel cathode (enhancing the passivity of the steel) and consumed at the anode. These and related reactions can carbonate the area around the anode (especially where carbon-based anodes are used, where the carbon also turns into carbon dioxide) and can lead to etching of the concrete surface and attack on the cement paste and even some aggregates once the alkalinity is consumed.

We can therefore see that three factors must be taken into account when controlling our cathodic protection system:

1   There must be sufficient current to overwhelm the anodic reactions and stop or severely reduce the corrosion rate.
2   The current must stay as low as possible to minimise the acidification around the anode and the attack on the anode for those that are consumed by the anodic reactions.
3   The steel should not exceed the hydrogen evolution potential, especially for prestressed steel to avoid hydrogen embrittlement.

The balancing of these requirements will be discussed below, under criteria for control of impressed current cathodic protection systems.

One of the more confusing facts of cathodic protection is that when we carry out a reference electrode potential survey of a reinforced concrete structure (Chapter 1) the most negative areas are those that are at highest risk of corrosion while the areas with a positive potential are at lowest risk of corrosion, i.e. cathodic. However, to achieve cathodic protection, we must depress the potential of the steel. This is because there must be excess electrons on the steel to force the production of hydroxyl ions (OH–) and suppress the formation of iron ions (Fe_2^+) which want to release electrons (equation (4.1)).

One of the earliest-identified criteria for achieving effective cathodic protection is to depress the potential of the cathodes to that of the most anodic areas (Mears and Brown, 1938). This stops the current flow from anode to cathode. It works because cathodes are more easily polarised than anodes. For a fixed current, an actively corroding area shifts its potential less than a non-corroding area. Therefore, once we can depress all the cathodes below the potential of the anodes, the corrosion will stop.

4.1.2 Impressed current cathodic protection system components

An impressed current cathodic protection system consists of an anode system in permanent contact with the concrete, an adjustable direct current power supply, reference electrodes embedded in the concrete near the steel and possibly other monitoring probes. There will also be a monitoring and control system to measure voltages and currents and to adjust the power supply. These components are all wired together. A structure will be broken down into zones which are powered, monitored and controlled independently of each other. The system is illustrated in Figure 4.2.

Impressed current cathodic protection anodes

The choice of anode is critical in the design of any cathodic protection system. For concrete there is a wide range of systems, each with their own advantages and limitations. Anode types currently widely available are as follows:

1   Mixed metal oxide coated titanium
   a. Expanded mesh (in a concrete overlay on the surface)
   b. Expanded mesh ribbon (in slots cut in the surface)
   c. Probe anodes (in holes drilled or cored into the concrete)
2   Conductive coatings
   a. Organic coatings (paint systems, typically chlorinated rubber or acrylics)
   b. Thermal sprayed metal (usually zinc but titanium has been used)
3   Conductive cementitious overlay (one proprietary system uses nickel coated carbon fibres in a wet sprayed mortar overlay, for example)
4   Conductive ceramic (non-stoichiometric titanium tube type probe anodes).

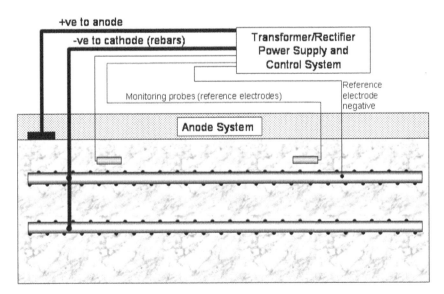

Figure 4.2 Schematic of an impressed current cathodic protection zone.

Table 4.1 Impressed current anode types

Anode	Relative life	Relative cost	Advantages	Limitations
Mixed metal oxide coated titanium mesh – embedded in a concrete overlay, usually sprayed	Very long	High	Durable	Overlay can debond if preparation and application is not done with great care and expertise Deadweight and thickness of overlay can affect structure
Mixed metal oxide coated titanium mesh ribbon	Very long	High	Durable No change in dead load or sections	Need adequate cover to avoid short circuit from anode to steel Leaves stripes on surface
Mixed metal oxide coated titanium probe anodes	Very long	High	Durable No change in dead load or sections	Titanium link wire can pit if voltage exceeds 8–12V. Anodes must be positioned to avoid short circuit from anode to steel Leaves stripes on surface Requires coring or percussive drilling to install
Conductive ceramic probe anodes	Very long	High	Durable No change in dead load or sections	Titanium link wire can pit if voltage exceeds 8–12V. Anodes must be positioned to avoid short circuit from anode to steel Leaves stripes on surface Requires coring or percussive drilling to install
Conductive organic coatings	Short	Low	Easy application and replacement No change in dead load or sections Hides repairs and givens choice of finish colour and texture	Needs dry conditions during application and will deteriorate rapidly in wet conditions Low resistance to wear and abrasion
Thermal sprayed zinc	Medium	Medium	Easy application and replacement No change in dead load or sections Moisture tolerant	Complex application process

A brief description is given in Table 4.1. For further information on anode systems see Broomfield (2007).

Monitoring probes

The reference electrodes are generally silver/silver chloride/potassium chloride electrodes or proprietary manganese/manganese dioxide electrodes designed and constructed for permanent embedment in concrete. They should be installed with minimal disruption to the concrete around the steel for representative measurement.

For systems on structures with very long lives, simpler, pseudo-reference electrodes may be used along with 'true' reference electrodes. This is because true reference electrodes have a life of about 20 years while a pseudo-reference electrode of graphite or mixed metal oxide coated titanium will last indefinitely but is less accurate.

Power supplies

The transformer/rectifier (T/R or rectifier) is the DC power supply that transforms mains AC to a lower voltage and rectifies it to DC. The positive terminal is connected to the anode and the negative to the cathode. The level of the output is controlled as described below. T/Rs can be run at constant voltage, constant current or constant potential (against a half cell). They can be adjusted manually, automatically by circuitry or computer control, or remotely using a telephone line and modem link or similar remote connection as described later.

Transformer/rectifiers for conventional cathodic protection systems of steel piles in docks or on pipelines can be very large and powerful, capable of delivering hundreds of amps, with oil-cooled transformers. However, for steel in concrete, the requirements are far more modest. Most systems are designed for a current density of about 10 to 20 mA per square metre of steel surface for actively corroding structures and for 0.2 to 2.0 mA/m^2 for new structures where there is no pitting and so no need to passivate pits.

Most steel in concrete needs less than 10 mA/m^2 to provide protection, usually at less than 10 V. The power for a 100 watt light bulb will typically protect 10,000 m^2. This means that a single phase, air cooled T/R will usually protect even the largest structure and power consumption is rarely an economic concern.

When calculating the current demand in a system, there may be a requirement for allowances to be made for the current flow to lower layers of steel as well as the outer, corroding layer. Calculations must also make allowances for the voltage drops down the connecting cables and anode strings.

The T/R must be rugged and reliable with minimal maintenance requirements. It should be easy to maintain with good instruction manuals, circuit diagrams for maintenance and easy access to fuses and other consumable and replaceable components.

There are two opposing directions of T/R design at the moment. The first is to make a simple, rugged reliable design with high-quality components. This is checked manually every one to six months and an annual 'service' carried out. The other is to attach a microprocessor and logger system that can monitor and control the system. This means that data can be collected remotely and in some designs the system can be adjusted without regular site visits, requiring only an annual inspection as long as local personnel carry out quick checks on the condition of the systems (loose wires, etc.).

The number of systems an organisation has in operation is one factor in choosing remote control. It becomes more cost effective to collect and review data without site visits as the number of cathodically protected structures increases. The sophistication of the client and his consultant is another factor.

The reliability of the microprocessor system has not been reported in the technical press; the author's first few remote control systems installed in 1986/7 worked until the structures were demolished, although most of those were installed inside buildings, in very benign environments. Some systems are comparatively simple and will only monitor on and off potentials, current and voltage. It is not possible to change the current or voltage settings remotely on some of these systems. With modern microprocessors and the internet it is possible to store data online, and send alarms by email or text message if the system malfunctions or exceeds defined limit values of current, voltage or reference electrode potential. Systems have been developed that will commission and operate a system according to BS EN12696, the British and European standard for cathodic protection of steel in concrete.

Design, installation, monitoring and maintenance

An impressed current cathodic protection system can be procured by a number of routes:

- Design, build and commission by an experienced contractor, usually with an independent design check with all responsibility on the contractor. The client must provide as much information as possible about the structure and the contractor may have to carry out investigations as part of the works.
- Detailed performance specification and outline design by the client's engineer with the contractor carrying out a detailed design. The client's engineer should have carried out most of the necessary investigations and testing prior to issuing the tender documents.
- Detailed design by the client's engineer with the contractor selecting products to meet the design requirements and installing the system,

often requiring approval from the client's engineer who retains design responsibility.

An impressed current cathodic protection system should be designed by a competent person with correct training and experience. In Europe a new standard BS EN 15257: 2006 'Cathodic protection – competence levels and certification of cathodic protection personnel' is now being implemented in the UK by courses run by the Institute of Corrosion and the Corrosion Prevention Association for levels 1 and 2. Certificates are awarded by the Institute of Corrosion with level 3 certificates awarded after assessment of experience.

The design of a cathodic protection system is described in BS EN 12696:2000. It requires following steps:

- The selection of anode type or types (more than one may be needed)
- Anode layout and the breaking down of the structure into zones
- Determination of current requirements by selection of a suitable current density on the steel and calculation of the steel surface area for the concrete components of the structure
- Design of reinforcement connections
- Design the cable layout (numbers, sizes, junction boxes, conduits, cable trays etc.)
- Location and number of embedded reference electrodes and other monitoring probes
- Power supply location, type, capacity, a.c. power provision and remote monitoring telephone line or other telecoms options
- Documentation – drawings, specifications, method statements, quality plan, operations and maintenance manual.

Once a contractor is approved with necessary Heath & Safety, Quality Plan and other necessary systems in place, the usual installation process is as follows:

1 Conduct concrete repairs while checking electrical continuity of the steel and making negative connections to the reinforcement.
2 Install reference electrodes and other monitoring probes.
3 Install the anode system or systems.
4 Install wiring system.
5 Install control and monitoring system and wire up the system.
6 Carry out pre-commissioning checks.
7 Energise the system.
8 Commission the system.
9 Provide operation and maintenance manual to client containing all necessary drawings, documents and manuals.

10 Operate the system for 12 months and supply a 12-month monitoring report.

11 Hand over system to client who appoints a suitably qualified and trained engineer to continue monitoring the system (this may be the client's own engineer, the contractor's engineer or an independent cathodic protection engineer).

Advantages and limitations of impressed current cathodic protection

The main reason for choosing to use an impressed current cathodic protection system as part of the repair and rehabilitation strategy for a structure is that it controls corrosion across the whole area where anodes are installed.

When carrying out repairs to corrosion-damaged reinforced concrete it is important to be aware of the 'ring anode' or 'incipient anode' effect. This is illustrated in Figure 4.3, 4.4a and 4.4b. The problem is that by repairing the corroding anode, we generate new anodes around the repair. This is especially prevalent in chloride-contaminated concrete where the higher moisture levels and the chloride content lower the electrical resistance of the corrosion cell and allow greater separation between anode and cathode.

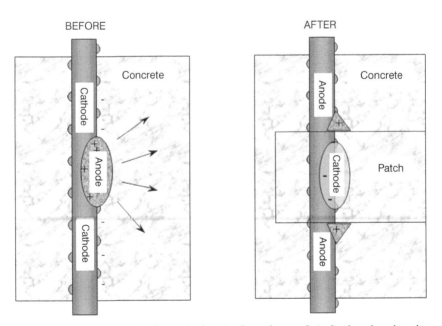

Figure 4.3 Incipient anode schematic showing how the anode is displaced to the edge of the repair by the formation of a new cathode in the patch repair (from Broomfield 2007).

Figure 4.4a Patch repair with surrounding spalling due to incipient anodes on a building (from Broomfield 2007).

Figure 4.4b Incipient anode showing repair (left side) and corrosion in original concrete (from Broomfield 2007).

Impressed current cathodic protection is frequently found to be the most cost-effective solution on reinforced concrete structures with chloride-induced corrosion with a long required life (over 10 years). However it is also widely used in other situations including carbonated structures and where any corrosion damage is unacceptable.

Another advantage of applying impressed current cathodic protection is that patch repairing can be made easier. As the corrosion protection is provided by the cathodic protection, cutting out and patching is easier as all chloride-contaminated (or carbonated) concrete does not have to be removed from behind the corroding reinforcement, as shown in Figure 4.5.

There are now well-proven national and international standards for supply, installation, monitoring and control of impressed current cathodic protection. These include the European Standard BS EN 12696 (2000), NACE standard RP 0290 (2000), and Australian/New Zealand standard AS 2832.5 (2002).

The key advantages of impressed current cathodic protection are:

- It controls corrosion in all areas where anodes are applied.
- It stops the 'ring anode' or 'incipient anode' effect.
- It simplifies repairs as there is no need to remove chloride-contaminated concrete. This can avoid requirements to prop the structure during repairs.
- There are good, well-recognised international standards for applying the technique.

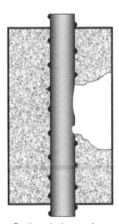 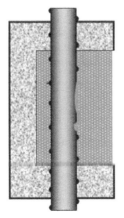 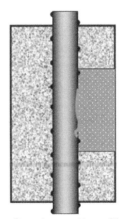

| Feathered edges and poor preparation allow breakaway at edges and poor keying | Squared edges cutting behind the bar and removal beyond the corroded area restores passivity and removes contamination | Concrete removal beyond the corroded area is not required for electrochemical treatment |

Figure 4.5 Patch repairs: bad, good and compatible with electrochemical treatments such as impressed current cathodic protection.

- There are good trade associations and learned societies with members skilled in the design, installation and operation of systems.
- There are certification schemes for engineers who design, install and operate the systems.
- It has a proven track record of over 20 years standing.
- Systems should last 20 to 50 years and can be designed for longer durability of the major components.

The key limitations of impressed current cathodic protection are:

- They require specialist knowledge.
- Extreme caution and specialist advice is required to apply impressed current cathodic protection to:
    - Structures containing prestressing steel
    - Structures with coated steel (epoxy or galvanised)
    - Structures that have been epoxy injected or have any other impediment to current reaching the steel.
- It requires a permanent power supply (and telephone line if it is remotely monitored).
- It requires ongoing monitoring and maintenance.
- Initial cost is high and requires 10 to 20 years remaining life to justify the life-cycle cost.

4.2 Galvanic (sacrificial anode) cathodic protection

Galvanic cathodic protection has been in existence far longer than impressed current cathodic protection but has only recently been applied to steel in concrete. This is because it relies on a small current generated by a corroding anode to protect the steel and the driving voltage is very small so that high-resistance concrete can reduce the current to inadequate levels to achieve protection.

4.2.1 Theory

The method used is to connect the steel to a sacrificial or galvanic anode such as zinc. This anode corrodes preferentially, liberating electrons with the same effect as the impressed current system, e.g.:

$$Zn \rightarrow Zn^{2+} + 2e^-$$

This system is illustrated schematically in Figure 4.6.

The same phenomenon is used in galvanising where a coating of zinc is applied over steel to corrode preferentially, protecting the steel. However, the main restriction on this system is that the zinc has a small driving voltage when coupled to steel. This is only a few hundred millivolts, and gets smaller

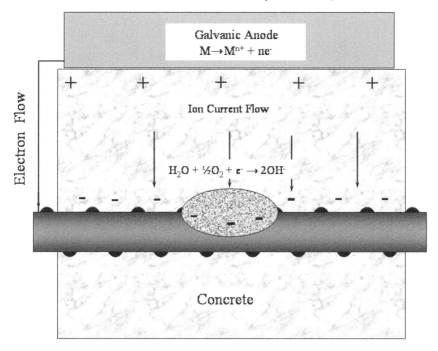

Figure 4.6 Schematic of a galvanic cathodic protection system. The steel is made cathodic by the dissolution of a suitable metal which preferentially corrodes, generating the electrons needed to sustain the cathodic reaction on the steel surface.

with actively corroding steel. While a galvanising system puts the two metals in direct contact, with galvanic anode cathodic protection there is an electrolyte to carry the current. The resistance of the electrolyte is crucial to the performance of the system.

Galvanic cathodic protection systems have been used extensively since the early 1990s in Florida on prestressed concrete bridge support piles in the sea. One of the reasons the galvanic system is used there is because concrete resistivity is low due to the marine exposure conditions. The Florida systems frequently incorporate a distributed anode of zinc fixed on the atmospherically exposed concrete and bulk zinc anodes in the water which pass current through the low-resistance sea water to protect the submerged area as shown in Figure 4.7.

The distributed anodes used in Florida are principally electric arc sprayed zinc, a few tenths of a millimetre thick (Figure 4.8), or zinc metal mesh mounted in a permanent form and grouted in place. The sprayed zinc anodes are also widely used in the USA for impressed current cathodic protection. The mesh anodes were originally mechanically clamped to the pile. In later versions a glass reinforced plastic (GRP) jacket containing the zinc mesh is

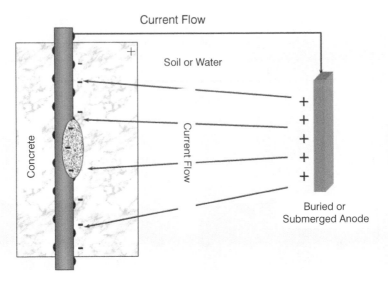

Figure 4.7 Schematic of a buried or submerged anode protecting reinforcement.

Figure 4.8 Thermal sprayed zinc being applied to a bridge substructure in the Florida Keys (courtesy Florida DoT).

attached to the pile and filled with grout after connecting the zinc to the reinforcement. All these systems and their performance in Florida are discussed in Kessler et al. (2002).

In the UK, the first major development was of a proprietary anode for patch repair. This has the appearance of a 'hockey puck' consisting of a disk of zinc in a specially formulated mortar that prevents the zinc passivating. Two pairs of wires protrude from opposite sides of the puck to be attached to the reinforcing steel exposed during the repair. By installing galvanic anodes within the repair (Figure 4.9) the 'incipient anode' effect is eliminated (see Figures 4.3 and 4.4).

The 'hockey puck' has recently been developed further into a small 'yogurt pot' sized anode with a single connecting wire that can be inserted into core holes in the concrete and wired together to produce a galvanic discrete anode system (Figure 4.10). They have been used predominantly on multi-storey car parks and on bridge substructures.

Since the maximum voltage that can be generated with zinc anodes is extremely unlikely to generate hydrogen embrittlement, galvanic systems have been used to protect prestressed concrete members. They are also used on fusion bonded epoxy coated steel reinforced piles as the effects of

Figure 4.9 Installation of a galvanic anode in a patch repair to prevent incipient anode induced corrosion around the repair (courtesy of Fosroc Ltd).

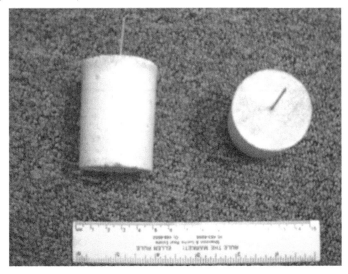

Figure 4.10 Galvanic anodes for installation in cored holes. Anodes are linked with titanium wires to form a zone.

electrical discontinuity between bars is unlikely to lead to significant stray current induced corrosion as the currents and potentials are low.

Table 4.2 summarises the different anodes and their characteristics:

- Thermal sprayed zinc – Zinc is flame or electric arc sprayed onto the concrete surface and a direct connection made to the reinforcement. It can be used as sprayed on marine splash or tidal applications. In dryer locations a humectant solution of hygroscopic salts can be applied. Over 50,000 m² has been applied, mainly in Florida.
- Thermal sprayed aluminium/zinc/indium – A proprietary variation on thermal sprayed zinc that uses an alloy to enhance the current rather than a humectant. A 300 micrometre thick coating is applied by arc spray. Over 35,000 m² has been applied.
- Adhesive zinc sheet – A proprietary system consisting of rolls of zinc 0.25 mm thick, 25 mm wide with a conductive gel adhesive on one side. The installed system can be painted. The edges must be sealed against water ingress as the gel can swell and leak.
- Encasement system – A proprietary system initially developed by Florida DoT consisting of expanded zinc mesh in a permanent form, grouted onto concrete piles or columns. Several hundred piles and columns have been protected with this system. It is for marine splash and tidal zone applications.
- Probe anodes in patch repairs – Proprietary anodes approximately 65 mm diameter by 30 mm high with four wires protruding for attachment to exposed reinforcing steel in the patch. Anodes are installed on a

Table 4.2 Galvanic anode types

Anode	Environment	Application	Durability/Life	Comments
Thermal sprayed zinc	Marine or anywhere (with humectant)	Requires bulky spray equipment and skilled operator	Greater than 10 y life expected. Reduces with severe exposure	Colour change to concrete is only effect
Thermal sprayed Al/ Zn/In	Marine and de-icing	Requires bulky spray equipment and skilled operator	10–15 y marine exposure, 15– 20 y in northern de-icing	Colour change to concrete is only effect
Adhesive zinc sheet	Not for very high wetting. Will work in very dry conditions	No special skills apart from soldering connections	Design life is 25 to 50 years. Gel deteriorates in very wet conditions	Anode is 1–2 mm thick. Either leaves zinc metal finish or painted metal finish
Encasement jacket	Marine only, mainly columns and piles	Special skills required for grouting up jacket	Very durable. Life up to 50 y	Pile column section enlarged by 25 mm or more. Load increase. Repair can be structural
Probe anodes in cored holes	Anywhere	No special skills	15 to 25 y life. Very durable	Core holes at 330 to 650 mm centres. Requires good design
Probe anodes in patch repairs	Any patch repairs	No special skills	10 to 15 y life. Very durable	Only area around patch repairs protected

maximum of 750 mm spacing around a patch. Approximately 200,000 sold protecting about 100,000 m^2 of concrete at the end of 2002.

- Probe anodes in cored holes – Potted-up zinc anode approximately 45 mm diameter by 40 or 60 mm long. Installed on 330 mm to 650 mm spacing in 50 mm diameter cored holes (other shapes and sizes are now becoming available)

4.2.2 The merits and limitations of galvanic cathodic protection

The principle advantage of galvanic cathodic protection is its simplicity, with no power required.

1 In principle this could lead to cost savings on a range of projects.
2 Its strength is also its weakness. The simplicity means there is no way of controlling it and there is no automatic flow of information

on its performance in the field. Monitoring systems can be installed if required, but at the expense of the simplicity of installation and operation.

3   Anodes have a limited life and need replacement at intervals that can be difficult to define. Most galvanic systems aim to provide at least ten years life before the anode needs replacing. However, systems have been designed to last up to 40 years.

4   If the concrete is too dry, the system will not work although under such conditions the risk of corrosion is likely to be reduced as long as the steel remains dry as well as the anodes.

5   Although well behind impressed current cathodic protection which has protected over 2 million m² of concrete structures, galvanic systems are progressing rapidly with at least 200,000 m² in less than five years of serious commercial exploitation.

6   There is now a range of anodes. These include:

- Thermal sprayed metals (usually zinc or alloys)
- Encasement systems
- Adhesive zinc sheet systems
- Embedded anodes for extending the lives of patch repairs
- Embedded anodes as discrete anodes.

7   The production of small galvanic anodes for patch repair systems is a useful adjunct to conventional patch repairs. It is not designed to provide full cathodic protection and would require a proper cathodic protection design to ensure adequate protection and a reasonable life of the patches.

8   The driving voltage of galvanic systems is low and may be inadequate to provide full cathodic protection in very high chloride conditions.

9   The limited driving voltage may make it attractive for protecting prestressed structures liable to hydrogen embrittlement and for epoxy coated reinforcement where electrical continuity cannot be guaranteed and the consequences of discontinuity are small.

4.3 Application of cathodic protection to highway and marine structures

The full range of anodes can be applied to highway bridge substructures. In the example shown in Figure 4.11, conductive organic paint coatings were considered to be too short-lived, and thermal sprayed zinc was offered as an alternative to conductive mesh plus overlay. It did not change the dead load or the profile and was a lower-cost option. However, the cantilevered ends were protected by probe anodes as they were more exposed to water run-down, as well the less accessible diaphragm walls between the longitudinal steel beams.

Figure 4.11 Thermal sprayed zinc applied to the leaf piers of Golden Fleece Interchange on the M6, UK. Probe anodes are applied to the cantilevered ends, the bearing shelf and the diaphragm walls between the longitudinal steel.

Figure 4.12 shows a series of beams and columns on a jetty being protected with mixed metal oxide coated titanium ribbon anodes. The cover was sufficiently thick to avoid short circuits and it avoided the increase in dead load, risk of overlay delamination and change in profile of a mesh plus overlay anode system.

4.4 Application to reinforced concrete buildings and car parks

Figures 4.13 and 4.14 show the application of conductive organic paint coating anodes to a reinforced concrete building with calcium chloride added to the mix as a set accelerator. The outer two bars and stirrups were corroding so the coating is only required on the outer face.

An example of conductive coatings applied to a car park soffit is shown in Figure 4.15.

4.5 Concluding remarks

Electrochemical techniques include impressed current cathodic protection, galvanic cathodic protection electrochemical chloride extraction and electrochemical realkalisation (see chapters 8 and 9). They treat a large area

Figure 4.12 Ribbon anodes being applied to the underside of a jetty in Jersey. Certain areas of the soffit, beams and all of the columns plus support trestles were protected using ribbon anodes. The tide rises to near soffit level and there is an 11 m tidal range.

Figure 4.13 Conductive coating anode being applied to a building (courtesy Taylor Woodrow Construction).

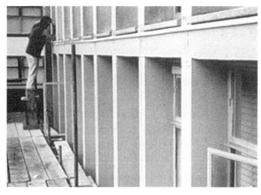

Figure 4.14 After application of the cosmetic top coat (courtesy Taylor Woodrow Construction).

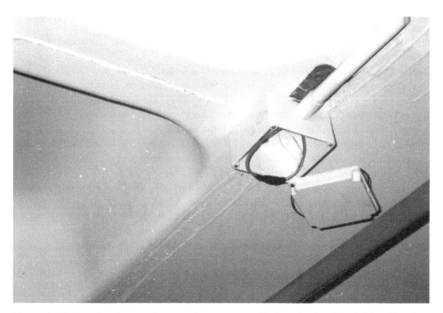

Figure 4.15 Conductive coating anode system applied to the 'waffle slab' soffit of an underground car park. The primary connections are via wires and tapes embedded in the concrete and collected at a junction box.

of a structure, not just patching up visually observable corroding areas. They do not give rise to 'incipient anode' problems (Figures 4.3 and 4.4, pp. 99–100). Concrete repair to damaged areas is less expensive and easier than for conventional patch repairs alone.

There are now comprehensive standards on impressed current cathodic protection for reinforced concrete and an impressive catalogue of case histories on a very wide range of structures. However, they require specialist knowledge from the designer and the installer.

They must be used with great care and careful selection on prestressed structures, those with ASR, those with epoxy coated rebars, epoxy injections or those with poor electrical continuity. A properly designed installed and operated system will provide long-term control of reinforcement corrosion.

Galvanic cathodic protection is a newer process but is expanding rapidly. It is simpler to install as it needs no power supply. However this makes monitoring more difficult and therefore it is harder to comply with the standards. The absence of monitoring makes it less expensive to run, but equally it is less clear when the system is no longer effective.

References

AS 2832.5 (2002) Australian Standard™ cathodic protection of metals. Part 5: Steel in concrete structures. Standards Australia, NSW Australia.

Broomfield, J.P. (2007) *Corrosion of Steel in Concrete: Understanding, Investigation and Repair* – 2nd edition. London: Taylor and Francis; 2007.

BS EN 12696 (2000) 'Cathodic Protection of Steel in Concrete', British Standards Institute, London.

BS EN 15257 (2006) 'Cathodic Protection – Competence Levels and Certification of Cathodic Protection Personnel'. British Standards Institute, London.

Kessler, R.J., Powers, R.G. and Lasa, L.R. (1995) 'Update on sacrificial anode cathodic protection of steel reinforced concrete structures in seawater', *Corrosion 95*, Paper 516, NACE International, Houston, TX.

Kessler, R.J., Powers, R.G., and Lasa, I.R. (2002) 'An update on the long term use of cathodic protection of steel reinforced concrete marine structures'. *NACE Corrosion 2002*. Paper No. 02254.

Mears, R.B. and Brown, R.H. (1938) 'A theory of cathodic protection', *Trans. Electrochemical Society*, 74, 519-531.

NACE (2001) *State-of-the-Art Report: Criteria for Cathodic Protection of Prestressed Concrete Structures*. NACE Report. 2002. NACE International. Houston, TX.

RP 02900 (2000) *Impressed Current Cathodic Protection of Reinforcing Steel in Atmospherically Exposed Concrete Structures*. NACE International. Houston, TX.

Bibliography

Broomfield, J.P. (2000) *The Principles and Practice of Galvanic Cathodic Protection for Reinforced Concrete Structures*. Corrosion Prevention Association. Technical Note 6.

Broomfield, J.P. and El-Belbol, S. (2006) *Impressed Current Anodes for the Cathodic Protection of Atmospherically Exposed Reinforced Concrete*. Corrosion Prevention Association, Technical Note 11.

NACE. *Sacrificial Cathodic Protection of Reinforced Concrete Elements, A State-of-the-Art Report*. NACE International. Houston, TX.

5 Cathodic protection using thermal sprayed metals

John P. Broomfield

Chapter 4 gave an overview of impressed current and galvanic cathodic protection. One of the most important choices in the design of a cathodic protection system is the anode system. Thermal sprayed metals are increasingly popular as anodes in both impressed current and galvanic anode systems.

5.1 Thermal sprayed zinc as an impressed current anode

Thermal spayed zinc was first used as an impressed current anode by the California Department of Transportation (Apostolos et al. 1987). The anode is applied using either flame spraying or electric arc spraying. The latter is now more common as deposition rates are far higher. There is a standard for thermal spraying anodes, see American Welding Society (2002). After initial surface preparation by a light grit blast, the zinc is sprayed onto the surface of the concrete using flame or electric arc equipment. Although the material is non-proprietary, it requires specialist applicators and bulky spray equipment.

As the coating is highly conductive it is important to avoid short circuits to the steel caused by tramp steel or tie wires. A simple electrical circuit with an audible alarm can be set up to warn of shorts between the reinforcement cage and the anode as it is applied; however, it can still be difficult to precisely locate short circuits. These must be eliminated for impressed current cathodic protection to work.

The anode is ideal for bridge substructures where it has been widely used in the USA. The coating is more tolerant of water after application than conductive paint anode systems. However, bulky specialist equipment is required for installation and it cannot be applied in any other way. The anode system has a typical life-expectancy of up to 25 years and has a medium to high cost of installation compared with other impressed current cathodic protection anodes.

Like a conductive paint there is little visual or other impact. The grey colour means it has limited acceptability on buildings. It can be overcoated, usually with silicate-based materials.

Due to its limited use in the UK and elsewhere in Europe there is limited home-grown information on its performance. However, based on the extent of US usage, there do not appear to be any major issues with its use. Hundreds of thousands of square metres have been applied to bridges in the USA.

The largest installations have been in Oregon where a number of historic landmark bridges were protected with this system (Covino et al. 2002). There have been other large installations on inland bridges elsewhere in the USA and Canada. Several hundreds of thousands of square metres of impressed current thermal sprayed zinc anodes have been applied to bridges in the USA. Figure 5.1 shows an impressed current cathodic protection installation on a bridge substructure in the UK.

5.2 Thermal sprayed zinc as a galvanic anode

The development of thermal spayed zinc as an impressed current anode was followed by its use as a galvanic anode by Florida Department of Transportation (Kessler and Powers, 1990) where thousands of marine bridge piles and columns have been protected. For marine applications it is

Figure 5.1 Thermal sprayed zinc applied to the leaf piers of Golden Fleece Interchange on the M6, UK. Probe anodes are applied to the cantilevered ends, the bearing shelf and the diaphragms between the longitudinal steel where moisture run-down might degrade the life of the zinc anode.

Figure 5.2 Thermal sprayed zinc being applied to a bridge substructure in the Florida Keys (courtesy Florida DoT).

recommended to use a bulk zinc anode at low tide level to protect the lower steel and to reduce the consumption of the zinc coating anode.

Many of the piles it has been applied to are prestressed. The use of galvanic zinc anodes means that the risk of hydrogen embrittlement of the prestressing is minimised.

The main advantage of thermal sprayed metal anodes are that they do not change the profile or dead load of the structure or require excavation of concrete. Most other galvanic anodes have at least one of these drawbacks.

While the thermal sprayed galvanic zinc anode was developed for application in a marine splash and tidal zone environment, it has been used inland in de-icing salt exposure conditions. In this case a humectant of hygroscopic salts was developed (Bennett, 1998). This was designed to reduce the electrical resistance of the cover concrete and the anode/concrete interface to boost the current flow of the anode in dryer inland conditions. A proprietary system with protective/cosmetic top coat is also offered in Europe.

In monitored galvanic anode bridge systems in Florida, sections of rebar have been isolated from the network and connected via an ammeter to the rest of the steel. This allows monitoring of protective current flows into a particular area. Similarly, areas of anode have been isolated and current flow measured to determine how much current the anode delivers. Occasional coring can be used to assess the amount of anode consumed and the anode replaced when necessary. Large-scale galvanic systems are in place protecting many piles in Florida where over 50,000 m^2 has been applied.

5.3 Aluminium–zinc–indium galvanic anodes

At about the same time as the humectant was under development, a proprietary aluminium–zinc–indium alloy thermal spray was developed (Funahashi and Young, 1999). This was aimed at boosting the galvanic current throw without the use of a humectant for inland de-icing salt applications. The original material was electric arc sprayed using a wire with a core of powdered metal. Later developments of other proprietary aluminium–zinc–indium anodes use a solid wire and are now widely offered in Europe and the Middle East.

5.4 Applying thermal sprayed metals

The anode is applied using either flame spraying or electric arc spraying. The latter is now more common as deposition rates are far higher than flame spraying. There is a standard for thermal spraying anodes on concrete, see American Welding Society (2002). After initial surface preparation by a light grit blast, the zinc or Al–Zn–In is sprayed onto the surface of the concrete using flame or electric arc equipment. The coating application rate and thickness must be carefully controlled as thick coatings can debond due to thermal shock of the hot 'splatters' of zinc (or Al alloy) hitting the cold surface. Coatings are usually a few tenths of a mm thick (typically 0.3 mm).

Zinc vapour can lead to 'zinc flu' if breathed in. Therefore all operatives must be suitably protected and the area shrouded to contain the zinc overspray.

This anode system has a life-expectancy of up to 25 years and has a medium to high cost of installation compared with other anode systems. Like a conductive paint there is little visual or other impact. The grey colour means it has limited acceptability on buildings.

References

American Welding Society. 'Specification for thermal spraying zinc anodes on reinforced concrete'. AWS/ANSI Standard. 2002; AWS C2.20/C2.20M:2002.

Apostolos, J. A. Parks, D. M., and Carello, R. A. 'Cathodic protection using metallized zinc'. *Materials Performance*. pp. 22–28. 1987.

Bennett, J. E. 'Chemical enhancement of metallized zinc anodes performance'. *Corrosion 98*. Mar 1998. Paper No. 640, San Diego CA..

Covino, B. S. Cramer, S. D. Bullards, S. J. Holcomb, G. R. Russell, J. H. Collins, W. K. Laylor, H. M., and Cryer, C. B. *Performance of Zinc Anodes for Cathodic Protection of Reinforced Concrete Bridges*. FHWA/Oregon DOT Report. FHWA-OR-RD-02-5. Mar 2002.

Funahashi, M. and Young, W. 'Three-year performance of aluminium alloy galvanic cathodic protection systems'. *NACE Corrosion 99* (Paper No 550) Apr 1999.

Kessler, R. J. and Powers, R. G. 'Zinc metalizing for galvanic cathodic protection of steel reinforced concrete in a marine environment'. *Corrosion 90*. Apr 23–27 1990; Paper 324: Las Vegas Nevada. NACE International, Houston, Texas.

6 Service life aspects of cathodic protection of concrete structures

Rob B. Polder and Willy H. A. Peelen

6.1 Introduction

Cathodic protection (CP) of concrete structures has been successful in stopping corrosion in reinforcement for over 20 years in Europe. Long-term experience has been reported by various authors (Tinnea & Cryer, 2008; Nerland et al., 2007; Wenk & Oberhänsli, 2007; Polder, 1998). Advocates usually state that the service life of CP is superior to that of conventional repairs. Moreover, CP installations involve significant investment and their service life is an important issue. Service life aspects of CP have not received much attention, however. Analogous to modern service life design methods for concrete structures (e.g. DuraCrete), a performance based, limit state oriented and probabilistic approach should be chosen. This will allow CP to be incorporated into modern maintenance management systems, which surely will be performance oriented. This may be an unreachable ideal at the moment; however, this chapter intends to present some of the necessary elements. Such an approach is based on identifying the most important failure mechanisms, modelling their time-dependency, quantifying uncertainties and finally, calculating failure probabilities as a function of time. This chapter describes time-dependent degradation processes and failure mechanisms of essential components of impressed current CP systems in the operation stage. The starting point is that the CP system has been properly designed and executed and that commissioning, testing and voltage/current adjustment in the early stages have been carried out successfully.

Good performance is defined as the absence of corrosion-induced damage. In practice, this is assumed to be the case if sufficient depolarisation is obtained (BS EN 12696). Sufficient depolarisation can only be reached when current flow is unhindered and control systems work properly. Normal maintenance should be applied, that is, two to four electrical checks and one visual inspection per year.

Degradation of the anode material itself, loss of concrete/overlay adhesion, corrosion of anode–copper connections and failure of reference electrodes are the most significant time-dependent failure phenomena. Quantitative models for degradation of the anode and overlay adhesion have been set up and validated from experimental evidence.

As part of a study into CP service life, an inventory was made of systems operating in The Netherlands in 2004. The companies who installed them were asked about maintenance, performance and failure of components. Information was obtained from a total of 70 CP systems installed between 1987 and 2002. Some were not very well documented, however. Data on 52 of them provided sufficient information to be used. In particular the evaluation of anode–copper connection and reference electrode performance was based on the field data obtained.

6.2 Degradation of anode systems

CP current flow from the anode to the concrete involves oxidising and acidifying electrochemical reactions at the anode/concrete interface. Their type and relative proportion depend on the availability of reactants, the electrical potential and the anode material. Their rate depends mainly on the current density. At the anode/concrete (or anode/mortar) interface hydroxyl ions are oxidised to oxygen gas by:

$$4 \text{ OH}^- \text{(aq)} \rightarrow O_2 \text{ (g)} + 2 \text{ H}_2\text{O(l)} + 4 \text{ e} \tag{6.1}$$

This reaction tends to lower the pH and is equivalent to acid production. Hydroxyl ions from the hardened cement paste tend to buffer a high pH. Acid production may eventually dissolve the cement paste.

When an oxidisable material such as carbon is present, it may be oxidised by:

$$C\text{(s)} + 2 \text{ H}_2\text{O(l)} \rightarrow CO_2 \text{ (g)} + 4 \text{ H}^+ \text{ (aq)} + 4 \text{ e} \tag{6.2}$$

or

$$C\text{(s)} + \text{ H}_2\text{O(l)} \rightarrow CO \text{ (g)} + 2 \text{ H}^+ \text{ (aq)} + 2 \text{ e} \tag{6.3}$$

In addition to consuming carbon, these reactions also produce acid and tend to lower the pH. Reaction (6.2) occurs at a lower potential than reactions (6.1) or (6.3), so thermodynamically it should be favoured. Experiments in solution at high pH have shown that reaction (6.2) represents only about 20% of the current and reaction (6.1) consumes about 80% (Eastwood et al. 1999), so kinetics apparently dominate the course of the reactions. The slow kinetics of reactions (6.2) and (6.3) under many different conditions are well known from energy related research in which electrochemical oxidation of carbon (coal) was studied.

Metals and chloride ions are potentially oxidisable. From the usual anode materials, titanium is strongly passivated (at normal potentials and pH) and the 'activating' noble metal oxides on the titanium surface cannot be oxidised any further. Oxidation of chloride is relatively small and can be neglected.

6.2.1 Oxidation of carbon-based materials

Carbon-based anode materials, such as carbon-filled conductive coatings, may suffer oxidising anodic attack (Brown & Tinnea; 1991). A tentative model can be based on Faraday's law, taking the current density at the anode/concrete interface into account, which may differ from the current density per unit surface area of concrete. For a surface covering conductive coating, the anode/concrete surface ratio is 1. A carbon oxidation efficiency factor *F(C)* with a value between 0 and 1 is introduced, that takes into account that not all current oxidises carbon:

$$\Delta m(C) = F(C) * i(A) * A(carbon) * z^{-1} * F^{-1} \tag{6.4}$$

with $\Delta m(C)$ amount of carbon oxidised per unit of time $[g/m^2/s]$
F(C) the carbon oxidation efficiency factor $[-]$, $0 < F(C) < 1$
i(A) the anode/concrete surface area current density $[A/m^2]$
$A(carbon)$ the atomic mass of carbon, 12 $[g/mole]$
z the number of electrons involved in reaction (6.2), 4 $[-]$
F Faraday's constant, 96,500 $[A \, s/mole]$.

For a surface covering carbon-based conductive coating CP system, a typical current density of 1 mA/m^2 of concrete surface area equals an anode current density of 1 mA/m^2. If the total current is oxidising carbon to carbon dioxide, so for F(C) = 1, this current density would oxidise 1.0 g of carbon per m^2 of anode surface area per year.

In more general terms, the amount of oxidation can be described with only one parameter, the oxidation efficiency factor *F(C)* and two input variables: the current density (assumed constant) and more conveniently written as I(A) in mA/m^2 and time. Filling in numerical values for constants in (6.4) and multiplying by time in years, we obtain:

$$\Delta M(C) = 1.0 * F(C) * I(A) * t \tag{6.5}$$

with $\Delta M(C)$ the total amount of carbon oxidised per anode surface $[g/m^2]$
F(C) the carbon oxidation efficiency factor $[-]$
I(A) the anode/concrete surface area current density $[mA/m^2]$
t the time [years].

For all the parameters listed above, time-averaged values should be used. Evidence from the literature to test this model is scarce. Observations on coating CP systems suggest a low level of attack (Polder et al., 2007). Samples from structures with a conductive coating (AHEAD from Protector) that had undergone CP for up to 9 years at about 1 mA/m² were studied by light microscopy (PFM) and scanning electron microscopy (SEM). The conductive coating had a thickness of about 150 μm, an estimated carbon content of 40% by volume (density 2250 kg/m³), totalling 135 g of carbon per m² of concrete surface. If it had been oxidised at 1 mA/m² for 9 years with $F(C) = 1$, about 9 g/m² of carbon would have been oxidised, so ca. 7% of all carbon in the coating. Oxidation would have occurred near the coating/concrete interface first, then progressively deeper into the coating, as illustrated in Figure 6.1. After 9 years, a zone of at least 10 μm thick near the interface would have become relatively devoid of carbon particles. SEM would probably have shown this, which did not seem to be the case. This would also have increased the electrical resistance of the CP system, which was not reported. The evidence suggests that not all of the current is involved in oxidising carbon particles, so $F(C)$ most likely had been significantly lower than 1.

Another source does not relate to a conductive coating CP system, but may provide useful information. Mietz and colleagues have reported data on samples taken from a CP system in Berlin after 15 years (Mietz et al., 2001). This system used carbon filled polymer cables as anode (FEREX from Raychem) of 8 mm diameter with a 2 mm copper core. A length of 10.7 running metres of cable was used per m² of concrete, so the anode/concrete surface ratio was about 0.25. During the first 7 years the system performed well, but it did not perform well during the remaining period. Increased system resistance required progressively increasing the driving voltage for sufficient protection (tested by depolarisation) in the later part of its life; at some point, sufficient depolarisation could not be reached any more. Autopsy of the anode cable was carried out using SEM and microprobe analysis after 15 years. It showed that carbon had disappeared from the outer layers of

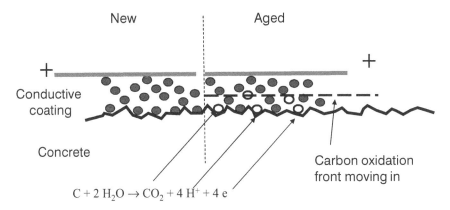

Figure 6.1 Model of oxidation of carbon particles in conductive coating anode; left: new material (grey symbols), right: aged material after oxidation of (white) carbon particles.

the polymer cables, down to a maximum of about 2 mm depth. The total carbon content (supposedly graphite plus polymer) was reduced from 90% down to 74% in the degraded outer layers. These changes had occurred over depths of typically 0.5 – 1.0 mm and the mean current density was 2 mA/m^2 (Mietz, personal communication, 2006). The amount of carbon oxidised in 15 years estimated from these analyses is equal to or greater than what corresponds to equation (6.5) with $F(C) = 1$. An explanation may be that the investigated samples had undergone a higher local current density than the average. Unfortunately, this renders these data unsuitable to obtain a value for $F(C)$. However, this case led us to the conceptual model of a carbon oxidation front gradually moving into the carbon-filled anode material, resulting in increased resistance.

Some other sources provide qualitative data. The performance of a Ferex system at Rotterdam (Schuten et al., 2007), was more favourable than that of the Berlin system. In Rotterdam, the anode system had not degraded significantly over 12 years of operation, as observed by light microscopy. Another Ferex system protecting 288 balconies at Capelle (NL) seemed to operate satisfactorily after 17 years (Nuiten, private communication, 2005). These cases may indicate that $F(C)$ can be well below 1.

A laboratory study (Eastwood et al., 1999) on carbon-filled polymer anodes (similar to Ferex) reported oxidation of carbon particles at high current efficiency in acid and neutral solutions ($F(C) = 0.8$ to 1). At high pH (> 11) and relatively high current densities, the carbon oxidation efficiency is low at about 20% and the remaining 80% of the current releases oxygen by reaction (6.1), so $F(C) = 0.2$. Most of the carbon seems to be oxidised to humic acids. Carbonate ions promote oxygen release and slow down carbon oxidation (at alkaline pH down to pH 9). Oxidation of carbon was found to increase the electrical resistance, very strongly at low pH and only moderately at high pH. Qualitatively, this corresponds well with the Berlin CP case. Taking into account that practical conditions may be more adverse than in the laboratory, we conclude that for the investigated anode material under normal CP operating conditions (a high pH, a relatively low anode potential and a low current density), $F(C)$ may be about 0.3. Of course this value may differ for different carbon-based anode materials.

Summarising, a conductive coating CP system can fail due to progressive oxidation of carbon particles from the anode/concrete interface. At some point in time, the electrical resistance of the coating layer that has become devoid of carbon particles is so high that sufficient current cannot be transferred to the concrete. Consequently, the system fails and the end of its working life has been reached. A similar type of failure has been reported for early carbon-based systems (Brown & Tinnea, 1991), particularly with high local current densities.

6.2.2 Limit state function for degradation of carbon-based conductive coatings

Our proposed model for oxidation of carbon in CP anodes takes into account the real contact anode/concrete surface area. An 'efficiency factor' is introduced to account for parallel reactions (6.1) and (6.2). This factor is quantified, from laboratory studies and practical evidence, to be about 0.3 for a carbon anode in alkaline concrete at normal CP current densities. As carbon oxidation proceeds the anode will degrade and eventually fail. The precise mechanism is not identified at the moment. The options include, for example, the decrease of the active surface area for oxidation according to reactions (6.1) to (6.3). Also carbon oxidation could occur in a small layer which becomes carbon depleted and whose thickness increases in time. As a consequence, the conduction between anode and concrete decreases in time. We opt for the latter mechanism as the most likely, based on microscopic observations. For carbon particle filled conductive coating anodes, a limit state function for oxidation induced degradation is introduced, Z_{ox}. Its resistance term R is the amount of carbon in a critical coating thickness in which carbon particles can be oxidised before the conduction becomes too low. Its load term S is the time and current dependent amount of oxidised carbon given by (6.5):

$$Z_{ox} = R - S \qquad (6.6)$$

with load $S = F(C) * I(A) * t$ from equation (6.5), and resistance R = mass of carbon in a critical thickness for conduction loss due to oxidation.

With $F(C) = 0.3$ and $I(A)$ known, S can be calculated. R must be estimated. In the Berlin CP system, after 15 years, a layer of $0.5 – 1.0$ mm (mean 0.8 mm) of the anode had become devoid of carbon (Mietz et al., 2001). After 7 years, the system stopped working properly. As conditions become increasingly aggressive (due to a lower pH and a higher anode potential), degradation may be non-linear with an increasing rate, so the critical thickness is equal to or less than $7/15 * 0.8 = 0.4$ mm. In about 7 years the system failed, so $Z_{ox} = 0$ and for $F(C) = 0.3$ and $I(A) = 8$ mA/m^2, the amount of oxidized carbon was about 17 g/m^2. Assuming that the same amount of oxidised carbon is critical for the AHEAD conductive coating, R is 17 g/m^2 and the critical thickness would be about 20 μm. Filling in these numerical values, the limit state function for oxidative degradation of a carbon-based coating anode is expressed in g of carbon per square metre as:

$$Z_{ox} = 17 – 0.3 * I(A) * t \qquad (6.7)$$

with $I(A)$ the anodic current density [mA/m^2] and t the time [year].

As an example, a conductive coating system at 1 mA/m^2 will have a service life of

$t = 17/0.3 = 57$ years. This seems quite long. It should be realised that this is a deterministic calculation based on mean values; the probability of failure after 57 years is ca. 50%. Uncertainties in the critical carbon mass, the oxidation efficiency factor and the (local) current density may cause earlier failure of anodes. Non-uniform current distribution would lead to earlier failure of high current density spots, which would increase the current density of the remaining active surface, which in turn would accelerate degradation. Increased local current density due to local lower electrical concrete resistance (for example due to water leakage) would definitely reduce anode service life, as can be observed in practice (see below).

6.2.3 *Loss of anode/(overlay)/concrete adhesion*

A conductive coating or a titanium/overlay system may fail due to acid attack of the adhesion with the concrete. Acid is formed at the anode by reactions (6.1) and (6.2). Acid formation and related failures, in particular due to local high current densities, have been reported (Brown & Tinnea, 1991). Adhesion is determined by the tensile strength of the concrete substrate, its surface preparation (cleaning, pre-wetting), the application method, the overlay materials' properties (shrinkage, E-modulus) and environmental actions (temperature and wetting/drying cycles), see for example, Julio et al., 2005. The requirements in concrete repair and/or CP standards are usually an average and/or minimum bond strength at 28 days age. Here it is assumed that materials and execution have been up to the requirements. During CP operation, the coating or overlay must be able to resist mechanical forces and environmental actions. Experience with cementitious overlays has shown that a bond strength of ca. 1 MPa is sufficient for a long overlay life. For CP overlays, EN 12696 requires an average of at least 1.5 MPa (minimum 1.0 MPa). For cementitious overlays in general, Dutch regulations require an average of at least 1.3 MPa (minimum 0.6 MPa). Such values can be obtained with proper surface preparation, materials and application. For coatings, a minimum of 1.5 MPa seems appropriate. These requirements are probably conservative and contain some reserve, even in the minimum values.

Here it is assumed that a bond strength of 0.5 MPa is sufficient to resist external action. So it is acceptable if some bond strength is lost from the original level of say 1.5 MPa. The difference between 1.5 MPa and 0.5 MPa is the capacity that can be consumed by CP-specific acid attack. A model is proposed for bond degradation due to CP-related acid formation. The acid dissolves parts of the cement matrix in the bond plane and the adhesion decreases. This model needs a description of the bond loss due to acid dissolution of the cement matrix and of the amount of acid produced for a unit of electrical charge.

6.2.4 Acid production

The CP current density must be converted into a rate of acid production. Possible electrode reactions and ionic migration and diffusion processes have been schematically represented in Figure 6.2 (Bertolini et al., 2004). They will be briefly discussed.

Electrolysis

The total amount of oxidation at the anode is the sum of all possible reactions: oxidation of hydroxyl, chloride and/or carbon. Only chloride oxidation does not form acid. Its contribution is probably small and can be neglected. Here it is simply assumed that all anodic processes consume hydroxyl ions. The 'flux' of hydroxyl consumption at the anode becomes equal to the total CP current divided by Faraday's number.

Migration

The current also causes mass transport through the concrete by ionic migration. Each ion carries a fraction of the current equal to its transport number $t(i)$. In aqueous solution comparable to concrete pore liquid (0.1 to 0.3 molar NaOH and/or KOH with a small amount of $Ca(OH)_2$), $t(OH)$ is 0.8 and $t(Na)$ is 0.2 (neglecting K and Ca). The migration fluxes of hydroxyl and sodium are given by the current multiplied by their transport number. Of the hydroxyl produced at the cathode, a fraction $t(OH)$ leaves the cathode area, that is $0.8 * I(CP)/F$. The excess negative charge at the cathode is compensated by inflow of an equivalent amount of positively charged sodium ions. The migration of hydroxyl ions will reduce the acid production at the anode. Some net consumption of hydroxyl will result anyway: an amount of sodium ions of $0.2 * I(CP)/F$ migrates from the

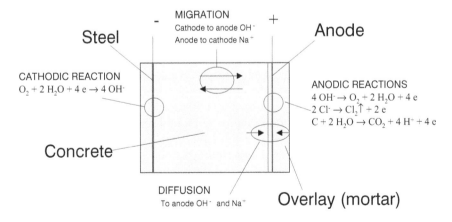

Figure 6.2 Schematic of electrode reactions, ionic migration and diffusion.

anode to the cathode. The overall effect of reactions and migration is that hydroxyl and sodium ions accumulate at the cathode and become depleted at the anode.

Diffusion

The sodium and hydroxyl gradients due to depletion and accumulation at the electrodes will cause diffusion of these species from elsewhere in the concrete towards the anode (and from the cathode to elsewhere). Diffusion is driven by a concentration gradient and not by electrical potential gradient. As a first approach, electrical potential differences are also assumed not to influence diffusion properties of the species involved. Consequently, the net acid production is the sum of the hydroxyl consumption mitigated by migration and diffusion of hydroxyl ions to the anode. The diffusion flux can be estimated from the diffusion coefficients of the ions in concrete and the concentration gradient near the anode. The NaOH diffusion flux, with a diffusion coefficient in the range of 1 to 10 * 10^{-12} m²/s and a concentration gradient of about 0.1 mole/l over a few centimetres (e.g. the cover depth), is about 10^{-8} mole/s.m². For normal CP current densities (1–10 mA/m²), diffusion has the same order of magnitude as ionic migration and the net acid production is close to zero. For significantly higher current densities the relative contribution of diffusion is smaller, so the net acid production will be higher.

The reduction of the OH consumption by migration and diffusion can be characterised by a net acid degradation factor, NADF, which is the ratio between the theoretical oxidation due to electrolysis at the anode J(acid,electrolysis) and the net oxidation reduced by migration and diffusion J(acid,net); the net acid production flux becomes:

$$J(\text{acid,net}) = J(\text{acid,electrolysis}) * \text{NADF} \tag{6.8}$$

Coupled migration and diffusion

Recently, new insights into transport of ionic species during CP have become available from numerical modelling based on coupling migration, diffusion and electrode reactions (Peelen et al., 2008). It appears that in the early life of the anode, steady state concentrations of all ionic species in the bulk of the concrete are obtained. This suggests that only transport of Ca and OH ions affects the acidification rate at the anode. Sodium and all other species not electrochemically active are merely re-distributed in the concrete until their gradients are such that migration and diffusion are in perfect balance (causing zero net transport). Further evaluation and validation of these numerical results are needed before quantitative results can be used with confidence.

Acid degradation model

Equation (6.8) gives the tentative model for acid production at the anode, which must be further quantified. Experimental evidence linking the electrical charge passed to acid dissolution in the anode/concrete interface is provided in (Polder et al., 2007, 2002). These papers describe microscopy observations on samples from conductive coating and activated titanium CP-systems, respectively, after several years of operation. In the coating study (Polder et al., 2007), acid attack was not observed by the two techniques used (PFM, SEM). Considering their resolution, this suggests that no more than between 1% and 10% of the total electrical charge had actually produced acid dissolution of concrete. Tentatively, it may be inferred that NADF was between 0.01 and 0.1. A similar range can be inferred from (Mussinelli et al., 1987). The titanium study (Polder et al., 2002) showed a quantifiable amount of acid dissolution after 5 or 6 years at 17 to 80 mA/m^2 titanium surface (4 to 20 mA/m^2 by concrete surface). The observed amount of dissolution was about one third of the amount expected from anodic oxidation and ionic migration, neglecting diffusion. Apparently, diffusion has compensated two-thirds of the acid formation. A fraction NADF of about 0.07 of the total charge had actually produced acid. This estimate fits well in the range of values derived from the coating study. Thus, the model for net acid production for current densities between 1 and 20 mA/m^2 is given by:

$$J(\text{acid,net}) = 0.07 * I(A) / F \tag{6.9}$$

with $J(\text{acid,net})$ in [mole/m^2 s].

With equation (6.9) we can calculate that for a CP current density of $I(A)$ mA/m^2, a net amount of $0.023* I(A)$ mole of acid per m^2 per year will be produced.

A recently developed numerical model was used to investigate acid dissolution at the anode; details can be found in (Peelen et al., 2008). Summarising, depletion of Ca(OH)^2 was found to progress with time, depending on current density and on the diffusion coefficients for hydroxyl and calcium ions. Diffusion coefficients were chosen on two levels that are thought to represent wet and semi-dry concrete, with a value for hydroxyl that is 10 times that of calcium. For high diffusion coefficients, the model calculations predict an acid degradation rate corresponding to a NADF of 0.063 for a current density of 1 mA/m^2 and 0.16 for 10 mA/m^2. This agrees quite well with our estimates from experimental evidence (0.07). At low diffusion rates, NADF could be as high as 0.78. This seems quite high, suggesting rapid acid attack that has not been observed in practice. Apparently, further work is needed, in particular on diffusion coefficients of ionic species used as input.

6.2.5 Loss of adhesion due to acid formation

The starting point for our model for loss of bond due to acid formation is a reduction of the bond strength from the required 1.5 MPa to 1/3, so 0.5 MPa, as a failure criterion. A macroscopically flat bond plane and homogeneous acid attack over the complete surface are assumed. The acid will dissolve an equivalent amount of cement hydration products. The bond capacity is expressed as a three-dimensional quantity, namely the amount of cement paste in a bond layer with finite thickness, see Figure 6.3. By definition, the cement hydrates outside the bond layer do not contribute to the bond. Considering the scale of roughness (interlocking) from microscopy, the bond layer may have a thickness of 50 μm. For 300 kg cement per m^3, a square metre of bond plane of 50 μm thickness contains cement hydrates equivalent to about 15 g unhydrated cement. The acid neutralisation capacity of cement is assumed to be equal to its CaO content, about 65% by mass for Portland cement. So the amount of acid that reduces the bond to 1/3 of its original value by dissolving 2/3 of the hydrated cement in a square metre of bond plane is equivalent to ca. 10 g of CaO, or 0.24 mole of acid (H+). This constitutes the resistance to acid-induced bond loss in the model.

6.2.6 Limit state function for CP anode adhesion loss due to acid production

For surface covering anodes (conductive coatings), the limit state function for current related acid formation induced bond loss is described as follows. The resistance R is the acid neutralisation capacity of a critical bond layer that may be dissolved before bond loss will occur. The load S is the acid produced as a function of time and CP current density, NADF * I(A) * t / F. For NADF = 0.07, a cement content of 300 kg/m^3 and a critical bond

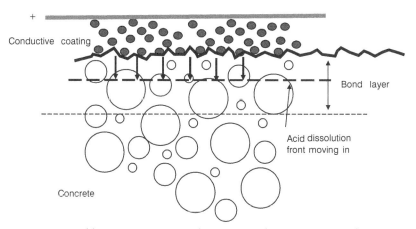

Figure 6.3 Bond layer in contact zone between conductive coating and concrete.

thickness of 50 μm, the limit state function for bond loss Z^b in mole of acid per square metre is:

$$Z_b = 0.24 - 0.023 * I(A) * t \tag{6.10}$$

with $I(A)$ current density [mA/m^2] and t time [years].

This expression applies for current densities between 1 and 20 mA/m^2.

As an example, for a current density of 1 mA/m^2 it would take ca. 10 years to reduce the bond strength to the criterion, from 1.5 to 0.5 MPa. This is unrealistically short: many coating CP systems in the field still have sufficient adhesion after (much) more than 10 years. This may be because the bond plane thickness is taken too low, the current density is lower than 1 mA/m^2, the bond criterion and/or the NADF of 0.07 from the titanium study are too pessimistic, or simply that the initial bond strength is higher. Other studies suggest that the reduction factor is between 0.1 and 0.01. The present model is apparently conservative, possibly by a factor of two. Further numerical modelling work may help to resolve this.

The model applies in homogeneous situations, which may not always be the case in practice. An inventory showed that local loss of bond occurred in some CP systems in less than 10 years. In at least one case, it was reported that the silicone joints between gallery slabs were leaking just above beams with CP that showed loss of coating bond. Apparently, moisture ingress had reduced electrical resistivity, increased current density and locally produced excessive amounts of acid.

6.3 Field performance of CP systems in The Netherlands

An inventory was carried out among companies in The Netherlands that installed CP to concrete structures. Object identification, year of installation, type of structure and protected area and anode type were recorded. Also they were asked about maintenance, repairs and failure of components. The inventory produced the following observations.

Between 1987 and ultimo 2004, about 75 CP systems were installed on various types of concrete structures like bridges, apartment and office blocks and parking garages, with a total protected area of 65,000 m^2 of concrete surface. This does not include about 40,000 m^2 of (small) precast ground floor systems (Schuten et al., 2001). Some systems were installed in multiple phases or had multiple anode systems. Most of the 75 cases are monitored regularly and work well. No corrosion-related damage has appeared as far as we know. Some systems have been neglected, however, and some cases have suffered from vandalism. 52 cases were well documented and had an age of at least 2 years. Their age distribution is shown in Figure 6.4 and their performance information is discussed in more detail.

Until 2004 there were 16 CP systems older than 10 years. Information on four of them is absent. One has been neglected and does not work properly

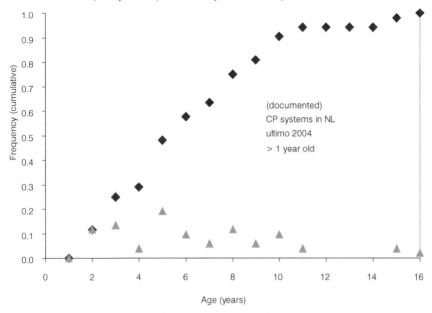

Figure 6.4 Age distribution of CP systems from inventory in The Netherlands; diamonds cumulative, triangles per year.

any more. Two of them have required moderate or major interventions. Nine operate satisfactorily. Out of eight systems aged 8 or 9 years, the performance of one is unknown and seven operate well. The youngest of these date from 1996, the year in which a CUR Technical Recommendation was published. So, these 24 oldest systems were designed and installed without a regulating document in place. Nevertheless, these old systems perform well: 16 out of 24 aged 8 to 16 years operate well without significant maintenance.

The inventory included 38 CP systems based on conductive coating anodes, aged 2 to 14 years. In six cases the anode failed to some extent, which was reported as:

- local disbonding (in one case at 3 years old, which may not have been working properly from the start; one after 9 years)
- poor condition, one at 6 years and one at 7 years old
- anode replacement, one at 7 and one at 10 years old.

In one case out of the total of 52 systems, anode–copper connections and reference electrodes failed after about ten years. This was the first CP project in The Netherlands, a bicycle path on a bridge installed in 1987. These items were repaired for a modest amount of money in 1999 (Schuten et al., 2007). Since then, it worked properly for at least several years (no recent information available). In a total of nine cases (including the bridge mentioned above), anode–copper connections failed at some point in time. This type of failure is discussed in more detail below. Summarising the inventory, in 33 out of 52 well-documented cases, the CP system worked well without failures or unforeseen maintenance. Figure 6.5 provides the probability of failure as a function of time, for anode failure, anode–copper connection failure and reference electrode failure. For anode failure, ranges are shown. The lower boundary represents the two cases of full anode replacement, the upper boundaries represent all cases including local anode disbonding.

6.4 Degradation of anode–copper connections

The connections between the primary anode (e.g. platinum wires or silver wire fabric used as primary anodes in coating systems) and the copper cabling are sensitive to accelerated corrosion. If the copper comes in ionic contact with the concrete, anodic polarisation dissolves the copper in a short time, causing loss of electrical connection. One option to protect a connection is by manually applying an isolating material around it, usually a polymer resin, like epoxy. The epoxy-encapsulated connection is then embedded in the concrete and the remaining space is filled with mortar (cementitious or epoxy). The absence of accelerated corrosion in the connection depends on the perfection of this isolation. Water ingress through the isolation provides a conduction path from the copper to the concrete pore system and promotes corrosion. Another option is to connect bare copper wires and primary anodes in connection strips inside a junction box, which is supposed to be watertight. In this case, water leakage into the junction box may provide a conduction path to the concrete and promote accelerated corrosion.

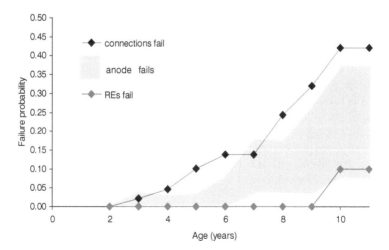

Figure 6.5 Probability of CP system failure as a function of time for anode failure (all related to conductive coatings, see text), anode–copper connection failure and reference electrode failure.

Practical experience has shown that embedded connections may fail. After about 10 years this had occurred in the CP system on a bridge from 1987 mentioned above (Schuten et al., 2007). In another case, connections between a series of precast façade elements were embedded. A significant fraction of the connections failed and had to be replaced. In other cases, embedded connections have functioned well for more than 10 years. Apparently the quality of the isolation, both from the 'design' of the connection and as executed on site, is important. In any case, embedded connections have a larger-than-zero probability of failure and consequently having as few as possible is beneficial. Similar experience exists with connections in junction boxes. In some cases, they function satisfactorily over many years; in other cases, leakage occurred and replacement was necessary. Locating junction boxes in sheltered positions near the façade (below gallery slabs) is beneficial rather than the exposed heads of cantilever beams.

From a reliability point of view, it is better to have multiple connections in parallel, rather than in series, where each individual connection failure may disconnect a large part of the protected area. Ideally, multiple connections should form a redundant network. This requires a large number of connections and cables, which is economically disadvantageous. From the inventory of Dutch CP systems it appeared that failure of connections occurred in about 2% of the systems at 3 years of age, increasing to 10% of cases aged 10 years or more. Figures 6.6 and 6.7 illustrate 'good' and 'bad' connection setup and positioning.

From the point of view of the service life of the CP system, avoiding corrosion of connections by careful design and execution, based on experience in the field, seems the best strategy. A test method for newly made embedded connections should be conceived, e.g. based on the electrical resistance between the anode cable and the concrete pore system.

6.5 Time-dependent failure of reference electrodes

Early papers on CP of concrete report frequent failures of embedded reference electrodes (REs). These cases probably reflect the growing pains of a new technology. From about 1990, several special types for concrete were well established, both silver/silver chloride and manganese dioxide types. However, due to their relatively high price, companies developed cheaper cells based on carbon or activated titanium. Again early failures occurred. Nowadays, cheap and reliable REs seem to be available. Causes of failure are known, such as shrinkage of embedding mortar and drying out of liquid-filled cells, causing loss of electrical contact.

From the TNO inventory, only one case out of 52 documented CP projects reported failure of REs after about 10 years, for cells dating back to 1987 (Schuten et al., 2007). However, more failures may have actually occurred. Cells may have been replaced as an unreported part of maintenance, or a

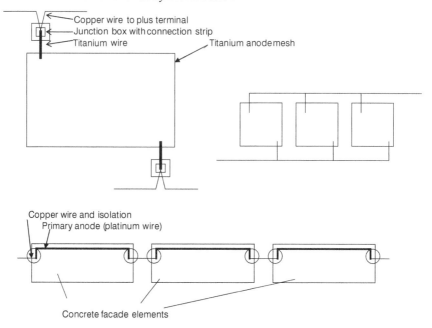

Figure 6.6 Multiple connections to individual CP system parts should be preferred (top) and series connections should be avoided (bottom).

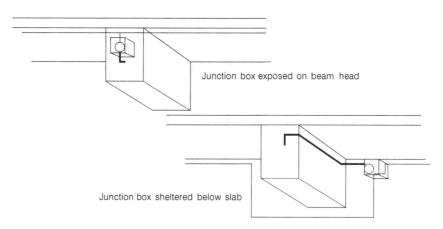

Figure 6.7 Junction boxes should be placed in a sheltered position (bottom) and not exposed, e.g. on beams heads (top).

few cells that did not function were simply ignored if the total number of REs was still sufficient for a good impression of depolarisation. Although the proper functioning of REs is critical from the point of view of depolarisation testing, their failure is not a big problem in practice. Either there is sufficient redundancy in the total number, or replacement is part of low-cost maintenance. As a strategy, having more REs than strictly necessary is a good investment. Tentatively, RE functioning may be checked by measuring their electrical resistance with respect to the reinforcement. In general, a rather narrow distribution of resistance values is found. A single value much higher than the rest may suggest loss of contact, indicating the RE is not working properly any more.

6.6 Summary and conclusions

This chapter describes time-dependent degradation processes and failure mechanisms of components of concrete CP systems in the operation stage. Most critical factors in time-dependent degradation of CP systems are oxidation of the anode material (carbon), bond loss of the anode or the cementitious overlay, corrosion of anode–copper connections and failure of reference electrodes.

Oxidation at the anode/concrete interface is inevitable. It has two effects: oxidation of the anode material itself, in particular carbon; and oxidation of hydroxyl ions. Both are important for the service life of concrete CP systems.

Oxidation of carbon particles in the anode material results in gradual build-up of a high electrical resistance and eventually the CP system fails due to insufficient current flow. The amount of carbon in a critical layer thickness within the anode is considered as the 'capacity' of the system. The 'load' is the amount of oxidation per unit of time and current density, which is only a fraction of the total current density. A limit state function was formulated with coefficients quantified from experimental data. No test method is available yet for determining the 'oxidation capacity' of a carbon-based anode material. Due to its chemical inertness, activated titanium does not suffer from oxidation. The service life of titanium-based CP systems is not determined by oxidation of the anode material itself.

Acid production in the anode/concrete interface results in dissolution of the cement paste and eventually loss of bond, which will block current flow and cause failure of the CP system. The processes involved are electrolysis, migration and diffusion. A model was conceived for acid-induced adhesion loss. The 'capacity' is a critical (bond) layer thickness of cement paste. The 'load' is the amount of acid formed per unit of electrical charge, which is a fraction of the total current density multiplied by time. A limit state function was formulated with coefficients quantified from experimental data supported by tentative numerical modelling. No test method is available yet for determining the 'acid capacity' of the anode/overlay/concrete in a CP system. Acid production induced bond loss is relevant for both carbon and titanium based CP systems.

Anode–copper connections in CP systems may corrode due to anodic polarisation if their isolation is less than perfect. If a connection fails due to corrosion, the part of the system fed by it stops receiving current and its corrosion protection will fail. Anode–copper connections are either isolated and embedded in the concrete or located in junction boxes. Their service life depends on the design and execution of the isolation against water penetration. A strategy of avoiding corrosion of connections seems appropriate, based on careful design and execution of connections, backed by experience. It seems possible to set up a test method for embedded connections. In practice, corrosion of connections occurs, but is felt to be a minor problem.

Reference electrodes in CP systems have failed, probably due to shrinkage and drying out, resulting in loss of electrical contact. Improved RE types and redundancy in their number per zone seem to be successful.

An inventory of CP systems in The Netherlands operating between 2 and 16 years showed that out of 52 well-documented systems, 33 operated without failure or unforeseen maintenance. The highest rate of failure reported related to conductive coating failure, ranging from local disbonding to complete replacement being necessary between 6 and 10 years, out of 38 conductive coating systems. Most of these failures were related to leakage, which causes current density to increase locally. Probably at those spots oxidation of carbon is accelerated, in accordance with our model. Nine cases were reported of failure of anode–copper connections due to corrosion of less than perfect isolation. This represents a significant amount of maintenance that should be avoided by using well-proven connection details, either embedded in the concrete or in junction boxes. Failure of reference electrodes was reported in only one case. However, more individual failures have probably occurred, but have been compensated by replacing failing electrodes or ignoring them. As reference electrodes are critical for evaluating system performance, having redundant numbers is preferred.

Overall, the inventory provides a positive picture of CP system performance over time. The highest failure rate, for conductive coating CP systems, suggests that avoiding (local) leakage is a critical factor; nevertheless, some systems will need local touching up or replacement of the anode after about 10 years.

Acknowledgements

The authors gratefully acknowledge fruitful discussions with the members of COST 534 working group Electrochemical Maintenance Methods and CP project information received from Dutch CP companies. Comments on a previous paper from which this chapter was developed (Polder et al. 2006) by Dr Jürgen Mietz of BAM, Berlin, are greatly appreciated. We thank

Mr Jan Leggedoor of Leggedoor Concrete Repair, Gasselternijveen, The Netherlands, for sponsoring the study on which this chapter is based. The project was supported by grant 040604.1183 from the Dutch Ministry of Economic Affairs in the TNO-co arrangement.

References

Bertolini, L., Elsener, B., Pedeferri, P., Polder, R.B., 2004, *Corrosion of Steel in Concrete: Prevention, Diagnosis, Repair*, Wiley-VCH, Weinheim.

Brown, R.P., Tinnea, J.S., 1991, Report on design problems for cathodic protection of reinforced concrete, Paper 122, NACE *Corrosion91*.

Eastwood, B.J., Christensen, P.A., Armstrong, R.D., Bates, N.R., 1999, Electrochemical oxidation of a carbon black loaded polymer electrode in aqueous electrolytes, *J. Solid State Electrochemistry*, 3, 179–186.

Julio, E., Branco, F., Silva, V., 2005, Concrete-to-concrete bond strength: influence of an epoxy-based bonding agent on a roughened substrate surface, *Magazine of Concrete Research*, 57 (8), 463–468.

Mietz, J., Fischer, J., Isecke, B., 2001, Cathodic protection of steel-reinforced concrete structures – results from 15 years' experience, *Materials Performance*, December, 22–26.

Mussinelli, G., Pedeferri, P., Tettamanti, M., 1987, The effect of current density on anode behaviour and on concrete in the anode region, 2nd Int. Conf. Deterioration and Repair of Reinforced Concrete in the Arabian Gulf, Bahrain.

Nerland, O.C.N, Eri, J., Grefstad, K.A., Vennesland, Ø., 2007, 18 years of cathodic protection of reinforced concrete structures in Norway – facts and figures from 162 installations, *Eurocorr07*, Freiburg (CD-ROM).

Peelen, W.H.A., Polder, R.B., 2004, Durability assessment of the concrete sheet piling for the new Dutch heavy-duty 'De Betuweroute' railway line, *Corrosion Prevention & Control*, March, 11–16.

Peelen W.H.A., Polder R.B., Redaelli E., Bertolini L., 2008, Qualitative model of concrete acidification due to cathodic protection, *Materials and Corrosion*, Vol. 59, 81–89.

Polder, R.B., 1998, Cathodic protection of reinforced concrete structures in the Netherlands – experience and developments, *HERON*, Vol. 43, no. 1, 3–14.

Polder, R.B., Peelen, W.H.A., Nijland, T., Bertolini, L., 2002, Acid formation in the anode/concrete interface of activated titanium cathodic protection systems for reinforced concrete and the implications for service life, ICC 15th Int. Corrosion Congress, Granada.

Polder, R.B., Krom, A., Peelen, W., Wessels, J., Leggedoor, J., 2006, Service life aspects of cathodic protection of concrete structures, *Concrete Solutions 2006*, 2nd International Conference on Concrete Repair, St. Malo, June 27–29, Eds. M.G. Granthem, R. Jauberthie, C. Lanos, 312–323.

Polder, R.B., Peelen, W., Leggedoor, J., Schuten, G., 2007, Microscopy study of the interface between concrete and conductive coating used as concrete CP anode, in M. Raupach, B. Elsener, R. Polder, J. Mietz (eds), *Corrosion of Reinforcement in Concrete, Mechanisms, Monitoring, Inhibitors and Rehabilitation Techniques*, European Federation of Corrosion, Woodhead Publishing, Cambridge, 277–287.

Schuten, G., Leggedoor, J., Polder, R.B., 2001, CP for corroding reinforcement in concrete floor elements, *Materials Performance*, Vol. 40, no. 1, 22–25.

Schuten, G., Leggedoor, J., Polder, R.B., Peelen, W., 2007, Cost aspects of cathodic protection; a review after 14 years of operation, in M. Raupach, B. Elsener, R. Polder, J. Mietz (eds), *Corrosion of Reinforcement in Concrete, Mechanisms, Monitoring, Inhibitors and Rehabilitation Techniques*, European Federation of Corrosion Publication number 38, Woodhead Publishing, Cambridge, 300–306.

Tinnea, J.S., Cryer, C.B., 2008, Corrosion control of Pacific coast reinforced concrete structures: a summary of 25 years experience, 1st International Conference on Heritage and Construction in Coastal and Marine Environment MEDACHS08, Portugal.

Wenk, F., Oberhänsli, D., 2007, Long-term experience with cathodic protection of reinforced concrete structures, *Eurocorr07*, Freiburg (CD-ROM).

7 Instrumentation and monitoring of structures

John P. Broomfield

Chapter 1 dealt with the different test methods available for assessing corrosion-related properties of reinforced concrete. Many of the techniques are applied with hand-held battery-operated equipment and a 'one off' reading is taken. In some cases such as cover meter measurements, that reading will not change. In others, such as concrete resistivity, corrosion rate or corrosion potential (reference electrode potential) it will change as aggressive agents in the environment move into the concrete and lead to or accelerate corrosion.

We can define corrosion monitoring as collecting corrosion-related data on a regular basis. In this chapter we will exclude strain gauge monitoring which is adequately covered elsewhere in the literature, and cathodic protection monitoring which is discussed in Chapter 4 and its references.

Corrosion monitoring can be done on new or existing structures. Its application to new structures requires forethought on the part of the designer, a clear understanding of what is required, what sensors and monitoring equipment is available and where to site sensors. Probe installation is relatively straightforward and can be integrated with the construction process along with power, signal and monitoring systems.

In existing structures corrosion monitoring may be by regular inspection using the techniques described in Chapter 1, or by probes installed in the concrete or on its surface. These will have wired or wireless connections to monitoring equipment.

There are two major issues with corrosion monitoring:

1 What can be monitored?
2 What are the merits of monitoring?

We can monitor with a limited number of techniques. The main techniques for structural monitoring are:

- Strain
- Crack/displacement
- Stress

- Acoustic emission
- Acceleration/vibration.

The main techniques for corrosion or concrete parameter monitoring are:

- Concrete resistivity
- Reference electrode potential
- Corrosion rate
- Linear polarisation resistance
- Galvanic current
- Relative humidity probes
- Temperature probes.

The reasons for carrying out monitoring are:

- Early warning of significant deterioration (limit state being reached)
- Input into planned maintenance
- Monitoring the effectiveness of repairs or protective measures being installed
- Development of a whole-life deterioration and costing model for a structure or group of structures.

7.1 Regular surveys to monitor corrosion

This has been done on motorway bridges in the UK where corrosion caused by de-icing salt leakage through expansion joints onto substructures is an acknowledged problem. A major cost is access on each survey as well as the manpower. The progress in chloride ingress and reference electrode potentials can be monitored along with a delamination survey.

Broomfield (2000) carried out regular surveys using linear polarisation resistance (LPR) on some reinforced concrete support pillars which had calcium chloride cast in as a set accelerator. These had been repaired with vapour phase corrosion inhibitors applied. Measurements from 1995 to 1999 showed the increase in the corrosion rate in treated and untreated control areas, with the corrosion rate peaking at around 0.5 to 1.0 $\mu A/cm^2$ (equivalent to a metal loss of 5 to 10 μm per year) with cracking seen approximately one year later. This demonstrated that the inhibitor applied was insufficient to deal with the chloride content of the structures. It also showed that a factor of two difference in corrosion rate between the treated and untreated areas made no significant difference to the rate and extent of cracking.

7.2 Permanent corrosion monitoring systems

A number of the techniques described in Chapter 1 are suitable for embedding in concrete for permanent monitoring. The main requirements are that they are durable when cast in concrete and exposed to the environment, the readings are meaningful and that readings do not drift with time as recalibration is difficult or impossible. By embedding probes and monitoring remotely we remove the access requirement and minimise the manpower needed. We also know that data were collected in exactly the same way and at exactly the same location for each successive set of data.

7.2.1 Permanent reference electrodes

The simplest probe is the reference electrode. These are designed for exposure in concrete for installation with cathodic protection systems (Chapter 4). However, it has been shown that once embedded in concrete, they cannot be recalibrated if they drift (Ansuini and Diamond 1994) and a very large number are required if a useful 'potential map' (Chapter 1) is to be produced. Reference electrodes are incorporated into LPR probes (next section) but are rarely used on their own except in a cathodic protection system where a potential shift is recorded so the absolute calibration of the electrode is not required.

7.2.2 Corrosion rate by polarisation resistance sensors

Corrosion rates can be measured by linear polarisation or by galvanic or macrocell techniques. Both have their merits and limitations as discussed in the next two sections.

The LPR probe is ideally suited to permanent installation, both in new construction and as a retrofit into existing structures. Probes consist of a reference electrode, an auxiliary electrode and a working electrode. In the case of a 'new build' probe, the working electrode can be a piece of steel of known surface area so that an accurate corrosion rate measurement is made on a working electrode in the same environment as the rest of the reinforcement. An electrical connection to the reinforcement means that measurements can also be taken on the reinforcement itself. Figure 7.1 is an example of such a probe installed in a reinforcing cage prior to casting.

Such a design cannot be used in existing structures. In such cases the working electrode must be the reinforcement in an undisturbed area of concrete. This means that the probe consists of a reference electrode and auxiliary electrode with a connection to the reinforcement at a suitable location. The probe assembly must be installed with minimum disturbance of the reinforcement, as shown in Figure 7.2.

7.2.3 Corrosion rate by galvanic sensors

A galvanic ladder probe for installation in new structures is shown in Figure 7.4. This consists of a series of mild steel anode rungs connected to a

Figure 7.1 A permanent LPR monitoring probe being installed in a precast deck unit of the Dartford Tunnel (Broomfield et al. 2003) (photograph courtesy BGB Projects Ltd).

Figure 7.2 A corrosion monitoring probe consisting of a reference electrode and an auxiliary electrode potted up in mortar in the process of installation into a continuously reinforced concrete pavement where corrosion monitoring is required due to the high level of chloride found in the mix water after laying several kilometres of concrete (Broomfield et al. 2003) (photograph courtesy BGB Projects Ltd).

Probe Measurement

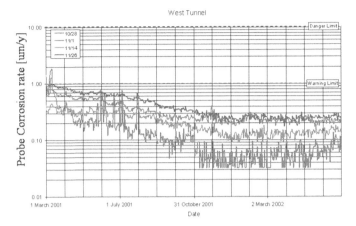

West Tunnel

Figure 7.3 Three years' data from corrosion rate probes shown in Figure 4.1 from the Dartford Tunnel, England (Broomfield et al. 2003) (courtesy NACE International).

Figure 7.4 Galvanic ladder probe (Raupach and Schiessl 1995). Ladders are at an incline through the cover so that successive rungs depassivate and a galvanic current starts to flow between the anode and a noble metal cathode (courtesy Dr Ing. M. Raupach).

stainless steel cathode via an ammeter. When chlorides or carbonation reach successive rungs the current between anode and cathode rises.

The ladder probe is designed for installation in new structures and has been installed in major bridges, tunnels and other structures thoughout Europe. Variations on this design are supplied by different manufacturers.

A more recent development is a 'washer probe' that can be fitted tightly into a cored hole to provide similar data on existing structures, as shown in Figure 7.5.

7.2.4 Corrosion rate by electrical resistance sensors

In an electrical resistance probe (ER), a strip, wire or tube sensor of known cross-section is exposed to the environment. In the case of concrete structures, it must be embedded at the time of construction otherwise it is not in the same environment as the reinforcement it is supposed to simulate. The sensor metal must be similar to the structure metal. The electrical resistance of the sensor is measured after initial installation and at subsequent time periods. As metal is lost by corrosion of the exposed surface of the sensor, the measured electrical resistance will increase which allows the amount of loss to be quantified. A proprietary automated corrosion probe reader is used to report the thickness loss directly. This uses a Wheatstone bridge

Figure 7.5 Galvanic 'washer' probe (Raupach and Schiessl 1995). The assembly is fitted into a cored hole in the concrete cover so that successive washers depassivate and a galvanic current starts to flow between the anode and a noble metal cathode (photograph courtesy Dr Ing. M. Raupach).

electrical circuit and uses an embedded thermometer to compensate for the effect of temperature on readings.

This technique measures losses from corrosion due to ineffective cathodic protection and is the only technique that can measure in-situ corrosion loss under cathodic protection. Valid measurements are possible even in non-conductive environments.

The main disadvantage of ER probes is that they only give valid data when the corrosion mode is uniform. These instruments are not suitable when corrosion is localised (pitting, cracking). This is its major limitation in concrete where corrosion initiates with pitting, particularly when chlorides are present. There is a trade-off between the sensitivity of the probe and its usable lifetime. The thinner the probe the more sensitive it is but the more quickly it is consumed.

Probes are permanently installed and measurements are made on a periodic basis. As a guide, the probe should be monitored at least once a month for the first 12 months. Thereafter, the monitoring frequency can be reduced to once every 3 months.

A graph is made of metal loss vs. time. The interval corrosion rate would be the slope of the graph between any two data points while the average corrosion rate would be the slope of the trendline as calculated by the least squares method. Many graphing programs, such as MS Excel®, have built-in capability to determine trendline slopes.

When used to demonstrate the effectiveness of a cathodic protection system, an average corrosion rate (trendline) of less than 0.1 mils/year (2.5 microns/year) over a 12-month monitoring period indicates that cathodic protection is effective at the location of the sensor.

Corrosion probes of the electrical resistance type are rarely used in concrete. They can suffer from localised pitting around the ends of the corrodible steel leading to rapid failure once corrosion initiates in this environment. They have been used to show that corrosion is under control, e.g. in impressed current cathodic protection systems, but in other cases have been found to be poor indicators of rates of corrosion.

7.2.5 Concrete resistivity sensors

Resistivity probes can be embedded in concrete. The design shown in Figure 7.6 was installed in the Dartford Tunnels on the River Thames estuary (Broomfield 2000). Data collected is shown in Figure 7.7.

7.2.6 Humidity monitoring

Commercially available relative humidity probes are available that are durable and are suitable for embedding in concrete section. The effect of relative humidity on corrosion rate is discussed in (Broomfield 2007).

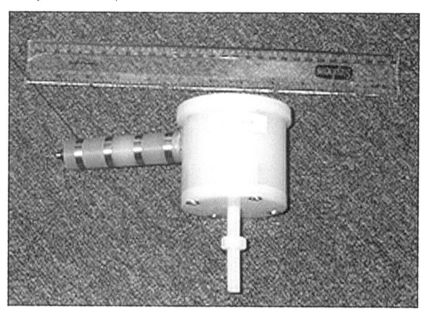

Figure 7.6 This resistivity probe is directly attached to the formwork so that when the concrete is cast around it the four stainless steel washers can be used to make four probe resistivity measurements.

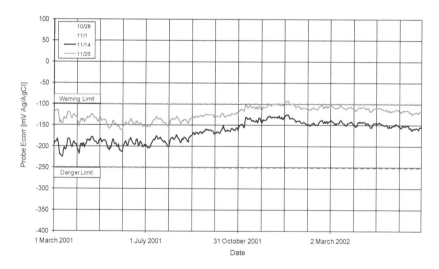

Figure 7.7 Over three years of data from resistivity probes (Broomfield et al. 2003).

7.2.7 Chloride content and pH monitoring

There is discussion in the literature about monitoring pH and chloride content (McCarter and Vennesland 2004, Buenfeld et al. 2008). However, there are no field data from such sensors and interpretation is difficult as the corrosion threshold is not a fixed value of pH or chloride content as discussed in (Broomfield 2007).

7.3 Remote monitoring systems and data management

There is a range of options for a corrosion monitoring system. For easily accessible systems with a limited number of probes, a simple termination box with sockets for plugging in meters is suitable. Figure 7.8 shows manual corrosion monitoring on a jetty with an LPR corrosion rate meter, a resistivity meter and a digital voltmeter for reading the reference electrode potentials (Broomfield 1998). The data in Figure 7.7 was collected manually.

Data can be collected automatically and stored either for download to a portable computer brought to site or by telecoms connection where the engineer can connect remotely to the computer and download data. The data in Figure 7.3 was collected on a central computer from networked sensors and downloaded remotely (Broomfield et al. 2003). Wireless systems are under development but are still unproven in the field at the time of

Figure 7.8 Termination box with connections for reference electrodes, LPR probes and resistivity probes on a jetty.

writing. An overall review of probes, data systems and data analysis is given in (Buenfeld et al. 2008).

References

Ansuini, F. J. and Dimond, J. R. (1994) Long term stability testing of reference electrodes for reinforced concrete. *Corrosion 94*. Paper 295.

Broomfield, J. P. (1998) Corrosion monitoring. *Concrete Engineering International*. Mar: 2(2): 27–30.

Broomfield, J. P. (2000) Results of long term monitoring of corrosion inhibitors applied to corroding reinforced concrete structures. *Corrosion* Mar. Paper No. 791. NACE Houston Texas.

Broomfield, J. P. (2007) *Corrosion of Steel in Concrete, Understanding, Investigation and Repair*, Taylor and Francis, London.

Broomfield, J., Davies, K., Hladky, K., and Noyce, P. (2003) Monitoring of reinforcement corrosion in concrete structures in the field. *NACE Corrosion/2003*. Mar: Proceedings CD ROM (Paper No. 03387).

Buenfeld, N. R., Davies, R. D., Karimi, A., Gilbertson, A. L. (2008) *Intelligent Monitoring of Concrete Structures*, CIRIA Report C661, CIRIA, London.

McCarter, W. J. and Vennesland, O. (2004) Sensor systems for use in reinforced concrete structures. *Construction and Building Materials*. Jul. 351–358.

Raupach, M. and Schiessl, P. (1995) Monitoring system for the penetration of chlorides, carbonation and the corrosion risk for the reinforcement. *Proceedings of the 6th International Conference on Structural Faults and Repair*. Feb. Vol 2 221–228.

8 Electrochemical chloride extraction

Ulrich Schneck

The target of electrochemical chloride extraction (ECE) is to reduce the chloride content in reinforced concrete non-destructively down to a level which is not critical for chloride induced corrosion activity. Within a short time (usually 4 to 8 weeks for single treatments) corrosion affected structural parts can be rehabilitated, and the corrosion protection of concrete for the embedded reinforcement can be re-established. This is different from cathodic protection (CP), which is meant to shift the reinforcement permanently into an immune state, where it cannot corrode regardless of environmental conditions. The feasibility and the efficiency of this treatment depend on many factors and can vary over concrete surfaces. Thus, they have to be evaluated thoroughly by an extensive condition survey and the experience of the ECE designer prior to an application.

8.1 Work principle

The effect of chloride removal is caused by an electrical field between the reinforcement and an external, non-permanent electrode (see Figure 8.1). This electrical field is controlled by the voltage between the electrodes, and all ions dissolved in the pore solution are moved – negatively charged ions such as chloride or hydroxyl ions towards the outside anode; positively charged ions (mainly sodium in case of de-icing salt attack) towards the reinforcement, which acts as cathode. The process requires wet concrete, and since the number of water molecules around cations is larger than around anions, more water will be moved into the concrete than out of it during such a treatment. The higher the voltage that can be set, the more intensive the chloride movement will be. Usually, 40 V is chosen for a good chloride extraction progress under still-safe work conditions.

The migration of ions is a physical process which is forced along the field lines between the electrodes: indeed it can happen only within the capillary and shrinkage pores, which take other directions than the established field lines. Due to the different size, specific movability (Elsener, 1990) and concentration of the dissolved ions, they obtain varying percentages of the total ion movement. Practically the portion of chloride in the anion

Principle of the electrochemical chloride extraction (ECE):

driving voltage between anode and cathode: ca. 30-40 V
max. current density, related to the rebar surface: ca. 1-5 A/m²
duration: ca. 4-8 weeks

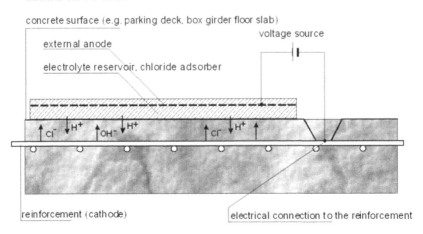

Figure 8.1 Principle of electrochemical chloride extraction (ECE).

movement is largest at the beginning of the ECE and decreases over the duration of the treatment, whereas the portion of hydroxyl ion migration is increasing at the same time.

Depending on the type of cement used for the concrete, some of the total chloride may be bound – mainly by the C_3A as Friedel's salt, but also by C_3S or C_4AF. The bound chloride does not cause corrosion activity. With the removal of free chloride from the pore area, bound chloride will be dissolved and gets free, since there is a dynamic relationship between free and bound chloride. The importance of this effect is related to the binding capacity of a concrete and is under discussion.

Electrochemical reactions take place on the reinforcement surface: the reduction of oxides, oxygen and water. All of them are related to the current and electrical charge which is impressed into the reinforcement/cathode by the ECE. Typical current densities (related to the reinforcement surface), range between 0.5 and 2 A/m², but can be much higher during the first hours/days of a treatment. Normally, the main process will be the reduction of oxygen and water, and according equation (8.1a) it will generate hydroxyl ions, which raises the alkalinity of the concrete in the reinforcement vicinity and is the main target of a related method – the electrochemical re-alkalisation:

Equation 8.1 Possible reduction reactions on the reinforcement surface forced by ECE.

$$\frac{1}{2}O_2 + H_2O + 2e^- \rightarrow 2OH^- \tag{1a}$$

$$\text{e.g.} \quad Fe_2O_3 + 3H_2O + 2e^- \rightarrow 2Fe(OH)_2 + 2OH^- \tag{1b}$$

$$2H_2O + 2e^- \rightarrow H_2 + 2OH^- \tag{1c}$$

Both processes – chloride migration and reduction of oxides, oxygen and water – run at the same time, but do not depend on each other in predictable terms. Whereas chloride migration depends on the applied voltage, the cover thickness and permeability of concrete and the water content, the current is controlled by the voltage, the temperature, the resistance of concrete (as a sum parameter of concrete cover, permeability, soluble ions and water content) and the charge transfer resistances on anode and reinforcement. This does not correlate with some other publications, e.g. (Polder et al., 1993), but has a practical backup when high amounts of chloride can be removed at quite low current densities and charges as well as only slow desalination progress being observed at high current densities and total charges.

So the ECE not only reduces the chloride content of the concrete but also raises its alkalinity as a result of the reduction reactions. This improves the corrosion protection additionally, since with a high OH– content also an increased chloride content can be present in the concrete without triggering reinforcement corrosion activity.

On the external anode, which is usually of a dimensionally stable material, we find other electrochemical reactions that lead to very acidic conditions: the oxidation of water (8.2a), hydroxyl ions (8.2b) and chloride as well as the formation of chlorine (8.2c). According to (Elsener et al., 1993) the reaction of water and chlorine can also cause an acidic environment (8.2d).

Equation 8.2 Possible oxidation reactions on the anode.

$$2H_2O \rightarrow O_2 + 4H^+ + 4e^- \tag{2a}$$

$$2OH^- \rightarrow \frac{1}{2}O_2 + H_2O + 2e^- \tag{2b}$$

$$2Cl^- \rightarrow Cl_2 + 2e^- \tag{2c}$$

$$H_2O + Cl_2 \rightarrow HCl + HClO \tag{2d}$$

8.2 Development, components and application systems

The idea of moving dissolved ions in solid materials with the help of an electrical field comes from geotechnology, where it is been used to clean soils. Between 1973 and 1975, ECE was tried first on a laboratory scale and then on a removed bridge deck in Ohio, USA (Slater et al., 1976, Morrison et al., 1976). With high voltages – up to 220 V – and developing high current densities, the chloride content could be reduced by more than 90% within a few days, but some drawbacks such as high temperatures, cracks and an increased permeability were caused in the concrete. Thus, the following ECE applications were done at reduced voltages and over several weeks.

In Norway, an application system that was developed and introduced in 1986 (Noteby, 1986) has gained the biggest portion of ECE applications and uses sprayed, wet cellulose fibres as an electrolyte reservoir, which provides an excellent connection between anode and concrete (see Figures 8.2 and 8.3). It is also known as NORCURE and can be used on large surfaces.

A very intensive research on ECE in laboratory and field scale was done within the Strategic Highway Research Program (SHRP) of the US Federal Highway Administration (FHWA) from 1988 to 1993 (Bennett et al., 1993b), which led to detailed recommendations and several practical applications in North America.

Generally, for an ECE the following components are needed:

- a dimensionally stable anode (DSA), usually activated titanium mesh
- an electrolyte reservoir that embeds and attaches the anode to the concrete surface
- a high power supply that establishes an electrical field between anode and reinforcement
- measuring and control units for recording and controlling voltage, current and – if present – the signals from installed reference electrodes.

Figure 8.2 Application of anode (here: mild steel) and sprayed cellulose fibres on a column of Burlington Skyway, Ontario, Canada in 1989. Photo courtesy John B. Miller.

Figure 8.3 View of a whole column of Burlington Skyway with partly applied electrode system. Photo courtesy John B. Miller.

Simple tap water can be used as an electrolyte. In order to prevent acidification of the electrolyte and of the concrete surface, frequently an alkaline solution made of $Ca(OH)_2$ or of $NaOH$ is used. Also from the SHRP comes the recommendation to take a lithium borate solution as electrolyte for a better alkaline buffering effect and to counteract expansive effects of ASR (Mietz, 1998).

Other application systems use geotextile blanket anodes with circulating electrolyte or tanks that contain the anode and an aqueous electrolyte (Broomfield, 2007). They can be mounted either horizontally or vertically.

A different technical solution was developed in 2001 in Germany by CITec and is designed for the focused treatment of smaller corrosion 'hot spots'. The basic layout can be seen in Figure 8.4 and has the following key features:

- The electrodes used for the ECE have a size of 60×60 cm, are pre-manufactured, re-usable and have an ion exchanger for binding chloride
- According to the configuration (rebar spacing, concrete cover, chloride content, concrete permeability etc) the electrodes can be combined in groups (max 10 m²).

- A chloride measuring unit (Schneck et al., 2006) signals saturation of the ion exchanger and switches off the related electrodes.
- The control of the ECE is based on a uniformly provided voltage of 40 V and a pulse width modulation (PWR) (Schneck et al., 2001) that switches the electrodes or groups of electrodes in intervals – normally 12 hours. The on-time is reduced when defined parameters such as current or potential readings are exceeded.
- Chloride saturated electrodes are regenerated in a solution at pH = 14. So the bound chloride is replaced by OH–, can be analysed to measure how much chloride has been removed and the electrodes get a fresh alkaline buffer capacity. Waste from the ECE application is thus avoided.

8.3 Influencing factors and side effects

As already mentioned, the effect of the ECE is influenced by many factors, which are not easy to determine in practical cases, and these factors can vary strongly within a few square metres. The ECE itself influences the treated structural part not only by the removal of chloride and increase of alkalinity, but also perhaps in unwanted ways if the structure is not assessed and the ECE is not designed and applied appropriately.

The reinforcement is the internal electrode for the ECE; concrete cover and rebar spacing have a great influence on the desalination effect. The main part of the process takes place between the concrete surface and the upper reinforcement layer; if the concrete cover is low – less than 20 mm – the area directly above the bars will be desalinated almost totally, but only a negligible chloride removal can be achieved in concrete areas behind the

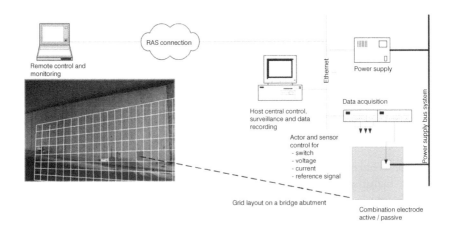

Figure 8.4 Schematic layout of the grid cell based ECE system by CITec.

upper reinforcement. Usually the chloride ingress has reached greater depths. Since the field lines get established from the anode mesh to the rebars, it is normal to find a rather patchy chloride removal in case of a widely spaced reinforcement layout; the desalination is concentrated then on the area of the rebars. If there is a high reinforcement density in the upper layer, it can shield the ECE effect from the second (deeper) reinforcement layer.

If either concrete cover or rebar spacing deviate within a treatment area, corresponding anode zones have to be defined. They should stay within 10 to 15 m². So the individual circumstances of the structure strongly determine how effective an ECE run will be and how much effort it needs.

Short circuits between reinforcement and anode – if the outside electrolyte can be in direct contact with rebars – must be avoided; cracks or visible reinforcement have to be sealed and tested before commissioning the ECE process.

8.4 Concrete composition and structural defects

The concrete composition, mainly the permeability – determined by cement type, cement content, w/c ratio, compaction, curing – also has a great effect on the ECE, as it influences the chloride ingress before treatment. Great care has to be taken with repair patches which can have much less permeability than the original concrete and hence limit the chloride migration.

Although it is recommended to repair all concrete defects before installing ECE, delaminated areas and cracks may be tolerated for the benefit of better site management in some cases, as long as these defects do not interrupt the chloride migration or result in short circuits. This has to be verified on a case-by-case base. If concrete repair has to be done before an ECE treatment, the repair concrete should have a similar quality in strength and permeability as the original concrete; if the repair layer thickness is large enough (> 4 cm), regular, unmodified concrete should be used.

8.5 State of corrosion at the start of treatment

If the reinforcement surface is entirely covered by corrosion products – which can be observed at high chloride concentrations in the reinforcement vicinity – the major initial reaction will be the reduction of oxides, which does not result necessarily in the formation of hydroxyl ions. Theoretically it can require up to 500 Ah/m² to reduce tightly covering corrosion products from the steel surface until oxygen reduction (and related hydroxyl ion formation) can predominate. This should to be considered at the definition of treatment targets.

This effect will be indicated by high initial currents and relatively low potential shifts measured by reference electrodes (considering IR correction). Investigations (Schneck, 1994) have provided the first indication of this effect and practical applications have verified this in many cases. Consequently, spray tests with pH indicators show interesting results on steel surfaces after ECE applications: titanium yellow, an indicator that shifts its colour from yellow to red within a pH range of 12 to 13 (which should be achieved when finishing ECE), changed its colour to red on rather blank reinforcement surfaces after low impressed charges from about 50 Ah/m², stayed yellow after applying 400 Ah/m² (see Figure 8.5) and became red also in visible pits after applying 1,200 Ah/m².

8.6 Alkali–silica reaction (ASR)

Alkali reactive siliceous aggregates show an expansive reaction in the presence of water and in a highly alkaline environment, forming cracks and alkali–silica gel. Thus, if such aggregates have to be used, a cement with low Na_2O equivalent should be used (or cement replacement materials such as PFA and

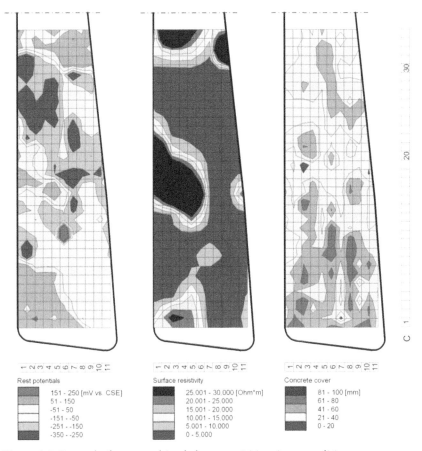

Rest potentials	Surface resistivity	Concrete cover
151 - 250 [mV vs. CSE]	25.001 - 30.000 [Ohm*m]	81 - 100 [mm]
51 - 150	20.001 - 25.000	61 - 80
-51 - 50	15.001 - 20.000	41 - 60
-151 - -50	10.001 - 15.000	21 - 40
-251 - -150	5.001 - 10.000	0 - 20
-350 - -250	0 - 5.000	

Figure 8.5 Example for a combined data acquisition in a condition survey – rest potentials, surface resistivity readings and minimum concrete cover.

GGBS can also mitigate the problem) and structures with ASR-endangered concrete are best kept dry. ECE can initiate or accelerate ASR because water is moved into the concrete – it can raise the water content up to about 8%, hydroxyl ions are generated by the oxygen and water reduction and sodium is moved into the concrete if NaCl is the origin of the chloride ingress, concentrating especially around the reinforcement used as the cathode.

Intensive research on this issue is available (e.g. Page and Yu, 1995), showing that no easy dependencies could be observed between impressed charge, time of application and effects on ASR, but it is advisable to pay attention to the presence of alkali reactive aggregates in the concrete of a structure subject to ECE, and if conditions seem to be critical, to run trial applications beforehand.

8.7 Chlorine evolution and acidification of the concrete surface

If no precautions are taken, the anodic processes according to equations (8.2a), (8.2c) and (8.2d) will generate large quantities of chlorine and very acidic conditions on the concrete surface. Chlorine is not only a health and environment hazard, but can establish a very corrosive atmosphere for neighbouring metallic parts (installation equipment, cars etc.), especially because the ECE requires wet conditions that lead to an increased air humidity adjacent to the application.

These effects can be limited by the use of DSA with high chlorine over-voltage (requires a special surface coating), alkaline and/or buffering electrolyte solutions or an anodic ion exchanger that binds chloride and releases hydroxyl ions at the same time.

8.8 Bond strength

The possible reduction of bond strength as a result of ECE has been investigated under various test conditions. Negative effects – the reduction of bond strength of up to 50% – have been found especially at smooth, corroded rebars after applying very high charges at high current densities, although no definite dependencies between the test parameters could be concluded. Vennesland et al. (1996) showed that an increase of bond strength was observed at applied charges of about 10,000 Ah/m² as well as a recovery of lost bond almost back to the initial value within one month after the termination of treatment.

Broomfield (2007) suggests that an increase of bond strength as a pre-stressing effect from corrosion products on the reinforcement has to be considered, which can be removed during ECE, but that there is no reduction of bond compared with the uncorroded state. Furthermore, if ribbed steels are used, the main bond is provided by the shape of the interface, so the possible slip on smooth surfaces cannot take place. Another influencing fact is the immense wetting of the concrete caused by the ECE. This might reduce the inner friction of the concrete (and the concrete–steel interface) and may also be responsible for a degradation of bond. As concrete dries out after finishing the ECE, this is a reversible effect which does not harm the concrete structure.

8.9 Use of ECE on pre-stressed concrete

A third possible result of the cathodic reaction is the evolution of hydrogen from water reduction (see equation (8.1c)). It can be triggered according to the Nernst equation at potentials (IR free) more negative than −770 mV vs. NHE (normal hydrogen electrode) or −1,070 mV vs. CSE (copper/copper-sulphate electrode) at a pH of 13 (which can be assumed on the steel surface). If evolved in larger quantities, hydrogen may migrate into micro-joints of pre-stressing steel and lead to hydrogen embrittlement and failure of pre-stressing steel. That is why ECE must not be applied on pre-stressed structures with immediate bond.

In post-tensioned structures using metal ducts, the pre-stressing steel is protected by tendon ducts that act as a Faraday cage. Unpublished studies by the author's company (Gruenzig, 2002) have shown that tendon ducts shield the duct inside even at larger defects (holes up to 3 cm^2) safely, so that in an aqueous solution an outside voltage of 40 V did not shift the potential inside the tendon duct more than − 50 mV. On the other hand, chloride cannot be removed with the help of electrochemical methods if it has entered the inside the duct. In reinforced concrete structures it is assumed that hydrogen cannot cause dangerous gas pressure on the steel surface, but diffuses through the capillary pores. A temporary reduction of breaking elongation of reinforcing steel has been observed (Bennett et al., 1993a) which recovered soon after stopping the ECE.

8.10 Initial condition survey

A solid, qualified condition survey is essential for any successful concrete repair. It provides the necessary information for the diagnosis and the understanding of the problems of a structure. The decision for corrosion rehabilitation by ECE has to be a result of all technological, financial and lifetime considerations and can be met in combination with other repair methods. The condition survey shall provide the following information:

- concrete cover and reinforcement layout (for evaluating the cathode surface)
- reinforcement continuity, possible short circuits to the anode
- state of the electrical connection between reinforcement and metallic parts within the possible treatment area (e.g. pipeline supporters, handrail fixings)

- extent of corrosion activity (detection of hot spots)
- state of corrosion on the reinforcement (may deviate from current corrosion activity!)
- extent of chloride ingress by size and depth
- profiles of chloride and water content down to ≥ 8 cm and behind the upper reinforcement layer
- concrete defects (spalls, cracks, delamination, patches).

It is obvious that all this information requires a sequence of different measurements – not only a potential survey, and the comparison of all values gives a very precise impression of the state and needs of a structure. This is demonstrated by Figure 8.5, where measurement results from a box girder floor slab can be seen. The rest potentials show some corrosion activity; the related chloride contents were up to 3.5% in the reinforcement vicinity. The surface distribution of the chlorides could be made visible by surface resistivity readings, which show a much larger extent than the corrosion activity. Chloride contents in locations with low surface resistivity (< 1,000 ohm.cm) and moderate potentials (ca. –150 mV vs. CSE) were about 0.5 to 1.5% in the vicinity of reinforcement. The concrete cover was quite high in the area with rather positive potentials and low in the areas with very negative potentials.

In conclusion, although the chloride content in the floor slab was generally high, it has resulted only in high corrosion activity where the reinforcement was getting closer to the surface. At the same time, corrosion active reinforcement was protecting reinforcement in less chloride-containing concrete. It was wrong to treat only the temporary corrosion active areas; chloride had to be removed in a larger scale.

8.11 Termination of treatment

It is very difficult to define, predict or to calculate parameters on which an ECE application can be terminated. As mentioned at the beginning, several factors influence the treatment progress and its dynamics. For a practical approach the following factors can be used:

- Chloride content: the average content of the surveyed profile should not exceed 0.4%, related to the cement mass; the maximum content in the reinforcement vicinity should be max. 0.8%.
- Impressed charge: depending on the extent of initial corrosion products on the reinforcement, the total impressed charge should range between 1,000 and 2,000 Ah/m^2, related to the reinforcement surface.

It has to be considered that the chloride content cannot be reduced to zero; often even an increase of chloride can be observed within the sampling depth during the treatment as chloride has penetrated far behind the upper reinforcement layer (see Figure 8.6). Then, a multi-stage application is required. The ratio/efficiency of chloride removal is less important than a generally low chloride content at the end.

Figure 8.6 shows how first the chloride within the concrete cover is mobilised, and as it is removed, more chloride arrives from deeper concrete zones. This may result in an increase of chloride content within the sampling profile and the (wrong) conclusion that the ECE might not have worked. If the amount of removed chloride can be analysed, this is very useful additional information to trace the ECE dynamics. Within a pause of about 6 weeks which is recommended to allow the concrete drying, more chloride is moved towards the concrete surface by capillary suction/evaporation. Furthermore, bound chloride can be released, but these are normally not large quantities. In a second treatment stage the chloride which has now accumulated in the concrete cover zone can be reduced to an uncritical level.

The impressed charge is the second parameter for the termination of the ECE; the above suggested range may be not achieved in case the reinforcement is still in good condition and the chloride content of the respective treatment area is rather low. Then, ca. 400 Ah/m² shall be a sufficient value, also because no or not much oxide has to be reduced on the rebar surface and hence almost all charge will be put into oxygen reduction.

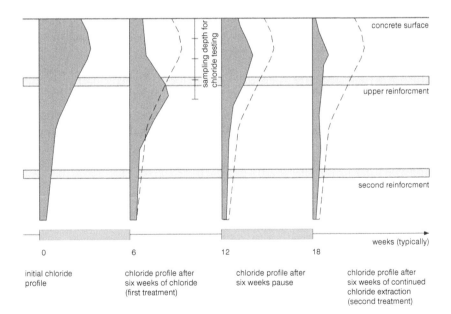

Figure 8.6 Dynamics of chloride migration during a multi-stage ECE application at high chloride ingress depths and with more than one reinforcement layer.

8.12 Other criteria of success

The total amount of removed chloride is another useful parameter; it depends on the total initial chloride content, so no values can be given for orientation. But in conjunction with the remaining chloride and the development of the chloride profiled during the application, the ECE progress and the necessary extent can be evaluated much better than referring only to the chloride in the concrete.

As stated above, the main task of ECE is to rehabilitate a concrete structure/element from corrosion activity. Thus, a repeated potential survey will show considerably different results: a generally more positive level of measurement values and a reduced spread between the maximum and minimum potentials. As macro elements are being dissolved by ECE, formerly very negative potentials have to become much more positive, and formerly positive potentials become more negative. Overall, the potential spread should stay within 150 mV. The repeated potential survey is not suitable to obtain the termination of treatment, because during ECE the reinforcement has been shifted deeply into the cathodic range. It has to recover from that polarisation back into a rest stage, and the high water content in the concrete has to be reduced to a normal level (from ca. 7% to 3%), until useful potential values can be measured. This requires at least 3 to 4 months.

Because of the wet concrete directly after the ECE, other repair actions especially the application of protective coatings, may have to be delayed.

8.13 Experiences from practical applications

Between 2001 to 2007, the author conducted ECE applications on 15 structures with treatment areas ranging between 10 and 280 m². Interest in ECE then waned in the UK in favour of cathodic protection, but is still used elsewhere, in Germany and the USA, for example. So far, more than 65 kg of chloride have been removed from over 1,200 m² of concrete, non-destructively, which corresponds to an amount of ca. 100 kg NaCl that had penetrated into the concrete. The elimination of corrosion activity could be verified in all cases. At the time of writing, more than 600 m² of concrete are under treatment. Based on the use of the CITec system, some typical projects and their results and conclusions are discussed in detail below, but see also (Schneck, 2006).

Box girder floor slab of a highway bridge

The bridge had corrosion problems resulting from leaks in the internal drainage system and some high chloride concentrations in the box girder floor slabs. A condition survey clearly determined the corrosion active areas, and since no significant structural damage was found, a non-destructive instant-off treatment was a useful approach, not least because the respective areas were above a high traffic bearing road and would have led to severe delays during a conventional repair.

The initial chloride profiles were varying in concentration and ingress depth; in some areas the entire cross-section of the floor slab showed chloride values > 2%, related to the cement mass. In a reference application on a 60 m² concrete surface, the effect of ECE has been investigated (see Figure 8.7), and despite more than 7 kg being removed, the chloride profiles down to 8 cm of concrete depth still contained chloride of more than 1.5% in places. A repeated potential survey 4 months after finishing the first treatment still showed corrosion active areas, so it was decided to run a second treatment, which was successful and eliminated all signs of corrosion activity.

Consequently, for the remaining five chloride containing corrosion active areas such a two-stage treatment was chosen from the beginning: 6 weeks ECE, 6 weeks pause and again 6 weeks ECE. In total, 25 kg of chloride was removed from ca. 200 m² concrete. The remaining chloride varied from 0.72% to 0.04% in the rebar vicinity (compared to more than 4% initially), but with an impressed charge of about 1,000 Ah/m² an appropriate alkalinity was developed to render even 0.72% as uncritical. (Corrosion depends on the ratio of chloride to hydroxyl ions, not just on chloride content.)

Figure 8.7 Layout of the anodes for the ECE treatment of the hollow box girder floor slab.

In Figure 8.8 a typical effect of a two-stage treatment of deeply chloride contaminated concrete can be seen; the initial chloride content is reduced in the outer 8 cm, but increases later, when more chloride is moved from deeper zones with capillary suction from drying processes. A second application is made for the final treatment and reduces the chloride content down to an uncritical level.

Five months after finishing the second treatment, a repeated potential survey showed a very even potential distribution with low deviation from the average value (–100 mV vs. CSE) and no signs of possible macro elements (and related corrosion activity) any more. The highest potential shift of about +400 mV has been observed in the areas of initially most negative potentials (Figure 8.9).

Centre column of a bridge across a highway

A very different target was set for this project – a column of a bridge over a highway close to Kassel in Germany had some chloride accumulated in the splash zone and needed to be rehabilitated preventatively by a combined ECE and hydrophobic gel treatment, so that an expensive and traffic disrupting conventional repair of this column could be avoided, and the long-term effect of a new hydrophobic gel application could be surveyed (Figure 8.10).

In this case, the effect of chloride reduction in the concrete was not that great, but was also not necessary yet. Another intention was to alkalise the reinforcement vicinity with respect to the following hydrophobic application.

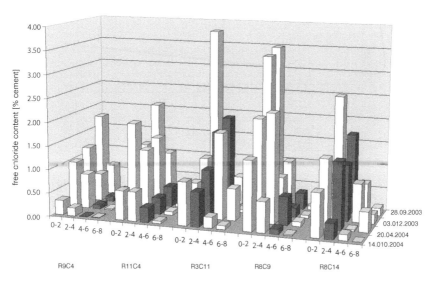

Figure 8.8 Development of chloride profiles in formerly high chloride containing locations; dark bars show chloride content in the vicinity of the upper reinforcement.

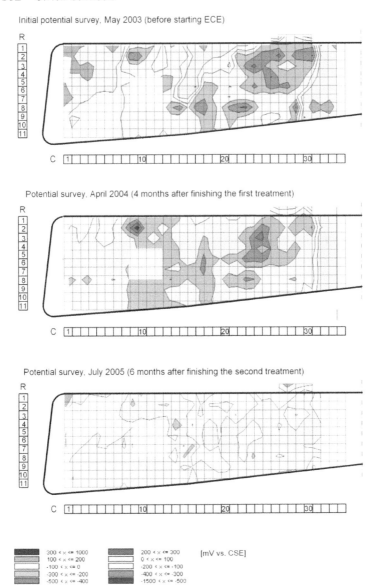

Figure 8.9 Potential readings before the ECE treatment, 4 months after and 6 months after a second ECE treatment.

Figure 8.10 View of the electrodes covering the splash zone of the column.

There, an increased carbonation was to be expected along with the drying of the concrete, and the alkalisation was done as a second preventive action. Figure 8.11 shows the achieved desalination effect after an application time of 6 weeks.

Bottom areas of columns and walls in an underground car park

The bottom areas of columns and walls in an underground car park in Bayreuth, Germany had accumulated large amounts of chloride. On the other hand, the loss of rebar diameter was still moderate, and no considerable signs of concrete deterioration were, as yet, to be seen. The chloride ingress was forced by capillary suction from the extremely wet car park slab into the dry walls and columns, where up to 11% of free chloride was measured.

To avoid expensive scaffolding support for the conventional repair of walls and columns and a long period of service disruption, ECE was chosen as an alternative, non-destructive repair. It was conducted over 8 weeks with only 2 days of restricted service of the car park (Figure 8.12). After its initial application, a successful treatment could be stated with more than 80% of the initial chloride content extracted from the concrete also behind the outer rebar layer, and in the ion exchanger of the electrodes 2.2 kg of chloride were measured – this equals an amount of 175 g/m² removed from the concrete.

Within 3 months after the treatment the chloride content went up especially in the outer 4 cm, as seen in Table 8.1. This effect could be related to the ventilation in the car park, which lowered the water content

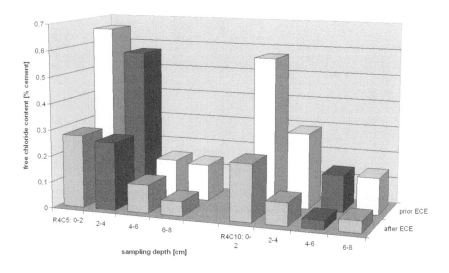

Figure 8.11 ECE effect from two sampling locations until 8 cm concrete depth (dark bars indicating reinforcement depth).

Figure 8.12 Electrode installation works at the column and wall section bottom areas.

Table 8.1 Chloride profiles (% of cement) before, directly after and 3 months after the ECE treatment; bold numbers from sampling depths within the outer reinforcement layer.

	Column	Wall section	Wall section	Column
Before the ECE				
0 – 2 cm		11.60	7.67	6.60
2 – 4 cm		4.74	1.90	3.35
4 – 6 cm		1.66	1.01	2.37
6 – 8 cm		1.17	0.55	1.97
Directly after the ECE				
0 – 2 cm	0.24	0.19	0.07	0.81
2 – 4 cm	0.22	0.56	0.22	0.89
4 – 6 cm	0.38	0.56	0.70	0.70
6 – 8 cm	1.17	1.27	0.96	0.71
3 months after finishing the ECE				
0 – 2 cm	0.38	0.93	0.19	0.20
2 – 4 cm	0.34	2.39	0.37	0.22
4 – 6 cm	0.68	2.54	0.97	0.58
6 – 8 cm	1.11	1.40	1.01	1.52

in the concrete from 4.5% directly after the ECE down to 1.8%. Strong evaporation and capillary suction moved more chloride from the floor slab into the treated area, as had led to the originally observed high chloride content.

It was necessary to change the ECE application strategy into either a two-stage treatment with the conventional repair of the floor slab prior or right after the first treatment or to use ECE as a supporting application for a cathodic protection of the floor slabs, which should bring down the excessive chloride contents to a moderate level and help to save effort for the instrumentation of the many protection zones that would otherwise be needed. In this case, the treatment time of ECE could be reduced to max. 2 weeks. A more detailed report about this project is to be found in Schneck (2005).

8.14 Conclusions

A thorough preparation is essential for a good ECE application, and its efficiency is variable; some questions remain, e.g. about practical parameters for the termination of an ECE treatment. Some general conclusions can be stated as follows:

- A detailed condition survey is very important for the correct configuration of an ECE application. It helps to focus the treatment on the relevant areas and to find criteria for planning treatment zones and application strategies.
- ECE is not limited to the space between concrete surface and upper reinforcement layer, but is effective as deep as reinforcement is present as a cathode. In most cases, chloride has also accumulated behind the upper reinforcement layer, so it can be removed by ECE. This requires multi-stage treatments with some weeks' pause between the applications.
- It is even possible to pull chloride through a 20 to 30 cm thick slab if the chloride containing concrete surface is not accessible. This requires at least a three-stage treatment.
- The ECE efficiency can be raised by using a pulse width modulation (PWM) mode which is more effective than permanent operation (Schneck et al., 2001). The chloride transport is forced along the field lines, but happens within the capillary pores that have other directions. So chloride can get caught in sack pores and be released during a switch-off period. The benefit of running ECE in a pulse mode has been verified recently (Elsener and Angst, 2007).
- In combination with the alkalisation coming from the reduction of oxygen and water on the reinforcement surface, some residual chloride can be accepted without ongoing corrosion activity.

The durability of an ECE depends on the concrete quality and the environmental influences on the structure after finishing the rehabilitation. If applied on unprotected bridge decks, 15 to 25 years are expected until new chloride-induced corrosion activity can be measured (Whitmore, 2001). If ongoing chloride ingress can be avoided or is not present (e.g. if a broken drainage system in a hollow box girder was repaired or concrete surfaces were coated), ECE will re-establish corrosion protection without limitations.

8.15 Fields of application

Quite different to cathodic protection (CP), ECE will only be used for rehabilitating corrosion cases. It is not feasible as a preventative action. The main application background is the treatment of corrosion caused by de-icing salts. Fire damage caused by PVC (which releases hydrochloric acid) is another possible case. Marine applications are less frequent, but have been done as well. Besides the technical feasibility, soft factors such as avoiding traffic disruptions, dust and noise or working in areas with limited access are important plus points.

In most cases, the treatment area of an ECE can be limited to 'hot spots' of a few square metres – ranging from ca. 5 to 100 m^2 (to be identified by the condition survey).

ECE can support other repair methods effectively and for a reduced total cost:

- Avoidance of expensive supporting frames for an alternative concrete replacement in load-bearing columns.
- Enlarging the protection sections/zones of a CP application by reducing the initial chloride content in locally deviating and very high chloride concentrations. So effort for the CP instrumentation and observation can be saved, especially when very small areas with high, deviating chloride content are to be considered.

In summary, ECE is an economic repair option for special cases which has more abilities than known from past experience and which should also be considered in combination with other methods.

8.16 Standardisation

Following a CEN technical specification on electrochemical re-alkalisation (CEN TS 14038-1, 2004), a new project on ECE is under progress at the time of writing, and is currently at the approval stage, as a Drafor Development (DD), (2007). NACE has published the Standard Practice paper SP0107-2007 'Electrochemical Realkalization and Chloride Extraction for Reinforced Concrete', which adds to the previously released state-of-the-art reports No. 24223 (2004) and No. 24214 (2001) about this topic.

References

Bennett, J., Schue, T., *Chloride Removal Implementation Guide.* SHRP-S-347, National Research Council, Washington DC, 1993a.

Bennett, J., Schue, T., Clear, K., Lankard, D., Hartt, W., Swiat, W., *Protection of Concrete Bridge Components: Field Trials.* SHRP-S-657, National Research Council, 1993b.

Broomfield, J., *Corrosion of Steel in Concrete – Understanding, Investigation and Repair.* Taylor and Francis, 2007.

Elsener, B., *Ionenmigration und elektrische Leitfähigkeit im Beton.* SIA Dokumentation D065: Korrosion und Korrosionsschutz, Part 5, 1990.

Elsener, B., Angst. U., Mechanism of electrochemical chloride removal. *Corrosion Science* 49, 2007.

Elsener, B., Molina, M., Böhni, H., Electrochemical removal of chlorides from reinforced concrete structures. *Werkstoffwissenschaften und Bausanierung* Vol. 420 Part 1 – Expert Verlag, 1993.

Gruenzig, H., Orientierende Versuche zur abschirmenden Wirkung eines Spannstahlhüllrohres im elektrischen Feld. CITec GmbH, 2002 (unpublished).

Mietz, J., *Electrochemical Rehabilitation Methods for Reinforced Concrete Structures – A State of the Art Report.* European Federation of Corrosion Reports 24, Institute of Materials, 1998.

Morrison, G., Virmani, P., Stratton, F., Gilliland, W., *Chloride Removal and Monomer Impregnation of Bridge Deck Concrete by Electro-Osmosis.* Report No FHWA-KS-RD-74-1, 1976.

Noteby, European Patent Application No 86302888.2, 1986.

Page, C.L., Yu, S., Potential effects of electrochemical desalination of concrete on alkali silica reaction. *Magazine for Concrete Research* 47, 1995.

Polder, R., Walker, R., *Chloride Removal from a Reinforced Concrete Quay Wall – Laboratory Tests.* TNO Report 93-BT-R1114, Delft, 1993.

Schneck, U., Zu Mechanismen der Stahlkorrosion in Beton bei der elektrochemischen Entsalzung. Dissertation, TU Dresden, 1994.

Schneck, U., Electrochemical chloride extraction as a substantial part of an optimized, cost and time effective repair of an underground car park. *Proc. EUROCORR 2005,* Lisbon, 2005.

Schneck, U., Experiences and conclusions from five years of chloride extraction applications. *Concrete Solutions, Proc. of the 2nd Intl. Conference,* BRE Press, 2006.

Schneck, U., Mucke, S., Gruenzig, H., Pulse Width Modulation (PWR) – Investigations for raising the efficiency of an electrochemical chloride extraction from reinforced concrete. *Proc. EUROCORR 2001,* Riva del Garda, 2001.

Schneck, U., Gruenzig, H., Vonau, W., Herrmann, S., Berthold, F., Chloride measuring unit for the improvement of security and performance of electrochemical chloride extraction. *Concrete Solutions, Proc. of the 2nd Intl. Conference,* BRE Press, 2006.

Slater, J., Lankard, D., Moreland, P., Electrochemical Removal Of Chlorides From Concrete Bridge Decks. *Materials Performance* 15 (11), 1976: 21–25.

Vennesland, Ø., Humstad, E., Gautefall, O., Nustad, G., Electrochemical Removal of Chlorides from Concrete – Effect on Bond Strength and Removal Efficiency. *Corrosion of Reinforcement in Concrete Construction,* Society of Chemical Industry, Cambridge, 1996.

Whitmore, D., *Guideline for Performance of Electrochemical Chloride Extraction to Concrete Structures.* AASHTO – FHWA, Washington DC, 2001.

9 Electrochemical realkalisation

Michael G. Grantham

9.1 Introduction: the mechanism of carbonation-induced corrosion of steel in concrete

Steel is normally passive when embedded in concrete, due to the alkalinity of the pore water. This passivity can be destroyed by carbonation of the concrete, that is, the gradual ingress of carbon dioxide into pores in the concrete, which, when dissolved in the pore water, forms mildly acidic carbonic acid.

This acid reacts with alkalis in the pore water to form a neutral salt and water, so the carbonic acid reacts with the calcium hydroxide to form calcium carbonate:

$$CO_{2+} H_2O \rightarrow H_2CO_3 \tag{9.1}$$

$$H_2CO_3 + Ca(OH)_2 \rightarrow CaCO_3 + 2H_2O \tag{9.2}$$

Once the calcium hydroxide is consumed the pH drops from 13 to 9 and the passive layer decays. The steel then corrodes in the presence of the oxygen and water available in the concrete pores.

Corrosion processes

When steel in concrete corrodes, it dissolves in the pore water and gives up electrons.

$$Fe \rightarrow Fe_2^+ + 2e^- \tag{9.3}$$

This is the anodic reaction.

The two electrons must be consumed elsewhere on the steel surface to preserve electrical neutrality:

$$2e^- + 2H_2O + O_2 \rightarrow 4OH^- \tag{9.4}$$

This is the cathodic reaction. Two interesting observations can be made here: more hydroxyl ions are generated in the cathodic reaction. These ions will strengthen the passive layer, warding off the effects of carbonation and chloride ions. It should also be noted that water and oxygen are needed at the cathode to allow corrosion to occur, so oxygen is not necessarily required at the corrosion site, provided the steel at the anodic site is electrically connected to a suitable cathode with access to oxygen and moisture.

So, if it is possible to make the system favour the cathodic reaction, corrosion will stop. In electrochemical techniques an external anode is applied to the concrete surface. This generates the electrons instead of the anodic reaction (9.3), and the steel has only the cathodic reaction (9.4) occurring on its surface. For both chloride removal (Chapter 8) and realkalisation, the external anode is temporary and the reactions are driven by a DC power supply. The systems re-establish a 'passive' environment around the steel that will last many years.

For all electrochemical treatment, good electrical continuity in the steel reinforcement is needed to ensure that current flows from the anode to all areas of steel. Electrical continuity must be checked and, if necessary, established by adding wired connections in all applications of these techniques. Equally, there must be no short circuits between the steel and the surface. If there are, current will short circuit the concrete pore structure and the chloride ions in chloride removal and the hydroxyl ions in realkalisation will not flow. Figure 9.1 shows a typical transformer rectifier for a temporary system such as is used for realkalisation.

Figure 9.1 Typical transformer rectifier used for electrochemical realkalisation and chloride removal (courtesy of Vector Corrosion Technologies).

Realkalisation

In equations (9.1) and (9.2) it can be seen how carbonic acid reacts with calcium hydroxide to form calcium carbonate. This removes the hydroxyl ions from solution, and the pH drops, so that the passive layer is no longer maintained and corrosion can be initiated.

The cathodic reaction (9.4) showed that by applying electrons to the steel, it is possible to generate new hydroxyl ions at the steel surface, regenerating the alkalinity, and restoring the pH.

The realkalisation process has been patented, and uses the same cassette shutter or sprayed cellulose system developed to apply chloride removal (though not the same as the CITec system introduced in Chapter 8). In addition to generating hydroxyl ions, the developers claim that by using a sodium carbonate electrolyte they make the treatment more resistant to further carbonation.

The patent claims that sodium carbonate will move into the concrete under electro-osmotic pressure. A certain amount will then react with further incoming carbon dioxide. The equilibrium is at 12.2% of 1 M sodium carbonate under atmospheric conditions.

$$Na_2CO_3 + CO_2 + H_2O \rightarrow 2NaHCO_3$$

The sodium hydrogen carbonate so formed is still significantly alkaline, and while sodium carbonate remains, can still buffer the pH so that it doesn't drop disastrously. In laboratory tests it has been shown that it is very difficult, if not impossible, for a treated specimen to carbonate again. Over 80 realkalisation treatments (50,000 m^2) have been undertaken on structures around Europe over recent years. These include the Hoover Building on the A40 into London and a number of large reinforced concrete hangars at RAF Wyton in Cambridgeshire. The treatment is faster than chloride removal, only requiring a few days of treatment. Work at Heriot Watt University (Al Khadimi and Banfill, 1996) showed that realkalisation actually improved the properties of concrete, reducing porosity and permeability, increasing strength and modulus and not apparently causing any difficulties. Figure 9.2 shows a schematic of a realkalisation system.

Anode types

Anode types are the same as for chloride removal. Cassette shutters or sprayed cellulose are used by the owners of the patented system, with a steel or coated titanium mesh. The steel is more likely to be used here as the treatment time is shorter and the steel is less likely to be completely consumed. Figure 9.3 shows a typical sprayed paper system and Figure 9.4 shows a typical cassette shutter system.

Realkalisation

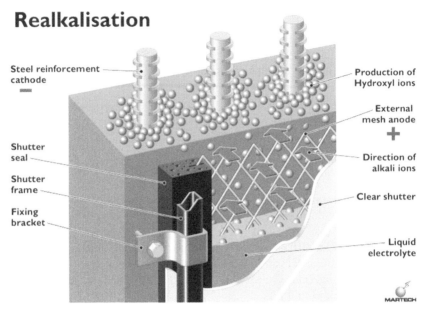

Steel reinforcement cathode

Shutter seal

Shutter frame

Fixing bracket

Production of Hydroxyl ions

External mesh anode

Direction of alkali ions

Clear shutter

Liquid electrolyte

MARTECH

Figure 9.2 Schematic showing the combined effect of hydroxyl ion generation at the steel and ingressing alkali solution combining to restore alkalinity in concrete (graphic courtesy of Martech Technical Services Ltd).

Figure 9.3 Sprayed paper system over mesh anode for realkalisation. The system uses a piped irrigation system to feed alkaline solution into the paper, while the anode and the reinforcement, acting as a cathode, are energised by a DC transformer rectifier.

Figure 9.4 Cassette shutter system used, in this case, for chloride removal on a bridge in Northamptonshire. A similar system can be used for realkalisation, though transparent plastic tanks are preferred to observe the anode condition and state of filling of the cassette tank.

Electrolytes

As stated above, sodium carbonate solution is the preferred electrolyte to give long-lasting protection against further CO_2 ingress, although potassium carbonate has been found to be less prone to leaching after treatment. However, introducing sodium or potassium ions can accelerate alkali–silica reaction (ASR) so in some cases just water is used, although this can evolve chlorine if any chloride is present in the concrete. Also, some problems have been found with adhesion of coatings when sodium carbonate is used, so the tendency is now to use a very low dose of sodium carbonate, or just to use water.

The UK Building Research Establishment comments:

Following the removal of the anode and thorough cleaning of the concrete, a protective (anti carbonation) coating is often applied. This will help to keep the concrete in a dry condition and to improve the aesthetics. A meticulous cleaning process must be followed to prevent crystallisation of electrolyte compounds behind coatings since this will lead to durability problems and coating failure. In line with good practice, integrity would benefit by the use of a fairing or scrape coat.

http://projects.bre.co.uk/rebarcorrosioncost/WebPages/
Realkalisation.html accessed 06/12/2010

Allowing the concrete to dry thoroughly after treatment and scrupulous cleaning is therefore another approach, with an appropriately formulated fairing coat perhaps able to hold back any crystalline salts which might disrupt coatings.

Typically the process uses an applied voltage of 10–40 V DC and a current of 1–2 A/m^2 of steel reinforcement. The voltage is adjusted to give the required current density. In such conditions, realkalisation treatment is usually complete in 5–7 days, or even faster if the concrete is very permeable.

It is easy to determine when the process is complete: a simple measurement of carbonation depth will show when it has been reduced to zero. Measurement of the ratio of voltage to current (effectively the resistance) is monitored and once it reaches a steady value, the process is deemed complete. Also, an on-site method for measuring sodium content of dust samples using a rapid ion selective electrode method, has been developed by the author (unpublished), where sodium carbonate is used.

As a smaller charge density is applied, the risks of damage are lower than for chloride removal. As mentioned above, ASR is a risk if sodium or potassium carbonate is used as the electrolyte.

Sodium carbonate can also cause short-term efflorescence, and the high alkalinity after treatment can attack some coatings. Sodium carbonate will attack oil-based paints, varnishes and natural wood finishes, but see the comments above on mitigating these problems.

9.2 Case histories

Well over 100 structures have been treated with realkalisation around Europe. Two cases are summarised above. Work has been carried out on a 1500 m^2 roof area of Walthamstow Magistrates Court. Carbonation depths ranged from 5 to 25 mm. After treatment a polymer modified mortar was applied to the surface and an elastomeric decorative finish was applied. The advantage to the Client in this case was the lack of noise, so the Court could remain in session.

What are the advantages of realkalisation over patch repair?

Realkalisation is a simpler and shorter treatment than chloride removal. However, the alternative of patching and coating with an anti-carbonation coating is much more effective than patching and coating for chloride attack. The extent to which carbonation has reached the rebar, and the requirements for patch repairing to restore alkalinity will determine whether realkalisation is preferred either because it is more economic, or in other cases to avoid the noise, dust and vibration required for extensive patch repairing. In practice, the technique comes into its own when the area to be treated is very large, when it is more economic to use a treatment that will deal with all the latent damage (areas which are carbonated but have not yet spalled) rather than individually identifying, breaking out and dealing with such areas. It is interesting to note that a very large number of contracts are using migratory corrosion inhibitors to deal with such latent damage, although the case for using such products is rather weak according to much of the research that has been carried out (see Chapter 10).

The technique has been used in Europe and the Middle East (in North America very little attention is paid to carbonation). The choice between realkalisation and patching and coating is a question of convenience and cost, together with a realistic appraisal of the effectiveness of anti-carbonation coatings.

The patentees and the licensees of the system claim that:

- It is financially competitive with the alternatives.
- There is greatly reduced vibration and noise.
- The entire surface is treated.
- Guarantees are offered.

9.3 Conclusions

Surprisingly, both realkalisation and chloride removal have struggled to gain a foothold, certainly in the UK, where cathodic protection seems to be the preferred option. In Germany, numerous contracts are using chloride removal, but research suggests that realkalisation is not often used. This is a pity for both methods as they are technically very sound and work very well.

Clients and Engineers should look again at this technology when dealing with large carbonated areas and UK contractors should take a further look at employing the technology in appropriate circumstances.

Acknowledgements

The author would like to acknowledge the help of Dr John Broomfield in preparing this chapter.

References

Al Khadimi, T. K. H., Banfill, P. F. G. et al., 'An experimental investigation into the effects of electrochemical re-alkalisation on concrete' *Proceedings of an International Symposium on the Corrosion of Reinforcement in Concrete Construction*, Cambridge. Royal Society of Chemistry. 1996.

Bibliography

Anderson, Gordon, Chloride extraction and realkalisation of concrete. *Hong Kong Contractor* pp 19–25, July–August (1992).

Broomfield, John P., SHRP Structures Research. Institute of Civil Enqineers ICE/SHRP, Sharing the Benefits; 29–31 October 1990; Tara Hotel, Kensington, London. London: Imprint, Hitchin, Herts; 1990; ICE 1990: pp. 35–46.

Mattila, J., 'Realkalisation of concrete by cement-based coatings.' Structural Engineering. Licentiates Thesis, Tampere, Tampere University of Technology, 1995.

Mietz, J., *Electrochemical Rehabilitation Methods for Reinforced Concrete Structures: a State of the Art Report*. Institute of Materials, London, 1998.

Yeih, W., Chang, J.J., 'A study on the efficiency of electrochemical realkalisation of carbonated concrete.' *Construction and Building Materials*, Elsevier, 2005.

10 Corrosion inhibitors

Michael G. Grantham

10.1 Introduction

Over recent years there has been a considerable amount of interest in the use of corrosion inhibitors in the repair of concrete.

There are two types of inhibitor: those mixed into a fresh concrete or repair at the time of placing, or those that penetrate by diffusion and/or capillary action through a hardened concrete from the surface. With the former type and careful mixing, a uniform concentration of inhibitor at the rebar can easily be achieved. With the latter, penetrating corrosion inhibitors, getting sufficient inhibitor to the surface of the rebars in sufficient quantity is the main difficulty, with only limited penetration likely with reasonably good-quality concrete that might nevertheless be contaminated with chloride salts. Penetration into low-quality concrete can usually be achieved, however, although this then leaves the question whether it can exit just as easily! In practice, therefore, application of coatings is usually recommended following a repair with a penetrating corrosion inhibitor.

Inhibitors can be classified into three main types:

- Inorganic, including nitrites, phosphates etc.
- Organic amines and some other organic compounds that include the presence of oxygen, nitrogen or sulfur atoms as well as the presence of double bonds. The lone-pair electrons of the mentioned atoms facilitate the adsorption process. These are adsorbed onto the metal surface and effectively block both the cathodic and anodic reactions.
- Volatile inhibitors such as amino-alcohols that have a high vapour pressure and penetrate into concrete in the vapour phase. Again, concrete with a low permeability which is chloride contaminated can be difficult to treat with such materials, unless the cover to the reinforcement is very low.

10.2 Types of inhibitor

10.2.1 Inhibitors mixed into fresh concrete or concrete repair materials

Calcium nitrite is the main material used for this application and has a long history of successful use, especially in the United States, with more than 300 structures using this inhibitor in a wide range of different types of structure (Page et al., 2000), provided the ratio of chloride to nitrite remains below about 1.8:1 (Broomfield, 2007, Vermani and Clemena, 1998). In practice this means a dosage of at least 22 kg per cubic metre of calcium nitrite dissolved in an appropriate amount of water (and that amount included in the total water content of the concrete when calculating the mix design). This assumes up to about 2% of cement as contaminating chloride and a cement content of, for example, 330 kg/m^3. Other researchers have suggested a lower nitrite to chloride ratio (Page et al., 2000) but caution needs to be exercised as too low a ratio can exacerbate pitting corrosion.

Calcium nitrite has been used in a good many structures to inhibit corrosion, some several decades old and a number of marine structures in the UK use calcium nitrite at least partly to inhibit corrosion. Information from Grace, one of the main suppliers, suggests more than 10 million cubic metres of concrete have been placed with calcium nitrite inhibitor worldwide. In practice, thousands of structures have undoubtedly been treated with either mixed-in inhibitors or by migratory systems.

There are other, organic, materials used in admixture with fresh concrete, such as amines, and esters which, in common with nitrites, do not retard the setting of the concrete, which can be a problem with some other materials. Nitrites, in fact, are set accelerators and are used with a retarder to combat rapid early setting.

The mechanism of the action of calcium nitrite as an anodic inhibitor depends on reinforcement of the passive film on steel, probably by oxidation of Fe^{2+} ion formed anodically at defects in the film.

$$Fe^{2+} + 2OH^- + 2NO_2^{2-} \rightarrow Fe_2O_3 + 2NO + H_2O$$

The Fe^{3+} ions so formed are stable and re-form the passive layer on the bar surface.

Since the 1990s, a number of proprietary organic inhibitors for use in concrete have been introduced, including various amines, alkanolamines, their salts with organic and inorganic acids and emulsified mixtures of esters, alcohols and amines (Page et al., 2000). None of these, to the author's knowledge, has as yet proved as effective as calcium nitrite, and with difficulties in establishing the chemical composition of these products, it is difficult to comment on their mechanism of action.

10.2.2 Inhibitors applied as surface treatments

Restoring alkalinity by surface treatments

Surface treatments for restoring alkalinity have been used, in which the treatment is applied to the surface of the concrete and hydroxyl ions penetrate the concrete surface by a combination of capillary action and diffusion. In practice, such treatments are likely to only be effective on rather permeable concretes, if the reinforcement is at any significant depth. Page et al. (2000) cites German and Finnish research into this method, and concludes that to have any chance of effectiveness, the coating would have to be quite thick, probably greater than 10 mm, to provide the required buffer of hydroxyl ions. Humidity would also be necessary to drive the diffusion of the hydroxyl ions into the concrete. In practice, this type of restoration of hydroxyl ions in the concrete is best done by electrochemical realkalisation, which has been proved to be effective (see Chapter 9).

Penetrating corrosion inhibitors

SODIUM MONOFLUOROPHOSPHATE

The American Strategic Highways Research Programme (SHRP) funded investigations of monofluorophosphate (MFP), and it was concluded that the material, when applied by ponding concrete with a 0.1 M solution, was not effectively transported into chloride-contaminated concrete and was thus ineffective (Page et al., 2000).

European reports on this product have been variable, with some workers finding that the material was adequately transported to normal cover depths after several applications by roller or spray application at 10–25% by mass or in solutions of approx 1–2 M. Other trials failed to show adequate penetration by the inhibitor, whether the concrete was pre-dried or not.

Schiegg and co-workers (Schiegg et al., 2000) carried out a three-year field trial involving both MFP fluoride and Sika Ferroguard inhibitors and concluded that neither material showed any significant effect on corrosion behaviour. Figure 10.1 presents three cumulative frequency graphs showing the behaviour of a reference panel and panels treated in 1997 with the two types of corrosion inhibitor, with a second line showing their behaviour two years later. The differences are insignificant.

On the other hand, Raharinaivo and Malric (1998) showed that it performed quite successfully in carbonated concrete.

CALCIUM NITRITE

While more normally used as an admixed inhibitor, some commercial patch repair products have included calcium nitrite, presumably in the hope that some of the nitrite will diffuse out of the repair into the surrounding concrete

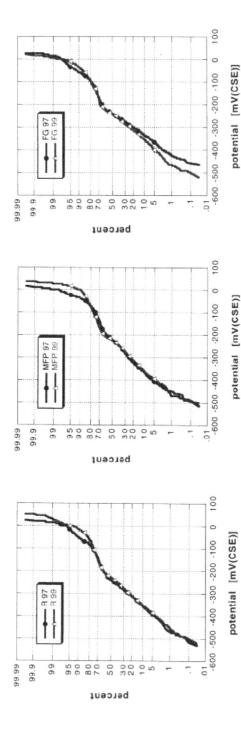

Figure 10.1 Behaviour of corrosion inhibitors (from Schiegg et al., 2000).

and thus might inhibit the incipient anode effect in chloride contaminated concrete. In the author's view, sacrificial galvanic anodes or a full CP system would be a better choice if any significant chloride content were present.

Cailleux, Pollet, Dubois and Michaux (Cailleux et al., 2008) also trialled the use of an ultrasonic pump to introduce a nitrite inhibitor into grout in tendon ducts, with some success. Figure 10.2 shows a photograph of the condition of rebars in grout with various levels of chloride after introduction of a nitrite inhibitor using an ultrasonic pump, after 12 months.

Sample autopsy after 2 months

Two months after treatment, the first autopsies were performed and half of the samples were fractured. The concrete covering and the ducts were removed in order to inspect the tendons. The Fig. 11 compares the corrosion state of the steel between the reference samples and the treated ones. The following observations can then be noticed:

- In the case of the chloride concentrations of 1% and 2%, a very clear reduction of the corroded areas was observed on the rebars after treatment. With 1% of chlorides, nearly no corrosion can be detected after injection.

- For the chloride concentration of 3%, corrosion layers can be observed both on the treated and the non-treated tendons. However, the treated samples might present a low corrosion state.

The presence of the inhibitor was evaluated by analytical test strips (Merckoquant 1.10020.000) which were applied directly at the surface of the tendons. The nitrite concentration was evaluated semi quantitatively by visual comparison of the reaction zone of the test strip with the field of a colour scale. The results showed a significant colour change for the strip applied on the surface of the treated samples indicating that the rebars were recovered by the nitrites after the treatment.

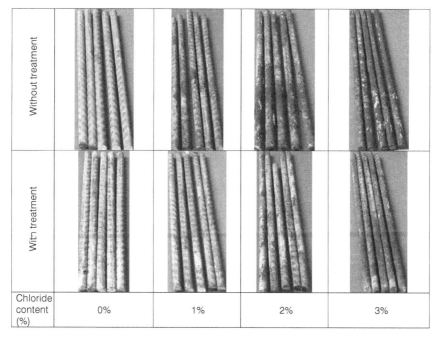

Figure 10.2 Comparison of the corrosion state of the treated and the non-treated rebars, one year after the injection of the inhibitor by an ultrasonic pump (from Cailleux et al., 2008).

PENETRATING VAPOUR PHASE INHIBITORS

These products, typified by Sika Ferroguard 903 and a number of other competitive materials, have been promoted for use on concrete suffering from reinforcement corrosion and often for concrete containing significant levels of chloride contamination.

Migrating corrosion inhibitors are typified by the amino alcohol based corrosion inhibitors, which coat the embedded steel with a monomolecular layer that keeps the chlorides ions away from the embedded steel. They also inhibit the reaction of oxygen and water at the cathodic sites on the steel which are an essential part of the corrosion process. As a result amino alcohols can be regarded as both anodic and cathodic inhibitors.

Thangavel and co-workers at the central electrochemical research institute in India (Thangavel et al., 2009) concluded that

- Alkalinity of the concretes was not affected by adding either MCI or admixed inhibitors. Among three concretes, Portland Slag Cements performed well under macro cell condition.
- Migrating inhibiting systems enhanced the passive condition of embedded steel in concrete by three times and admixed inhibiting systems enhanced the passive condition by two times when compared with the respective control concretes.
- Migrating inhibiting systems showed a 60% reduction in macrocell current, whereas admixed inhibiting systems showed only a 40% reduction in macrocell current in the case of PSC concretes.
- The prolonged passive condition of embedded steel, lowest macrocell and integrated current, lower free chloride contents, lowest corrosion rate, and visual observation data confirmed the better corrosion resistance of steel embedded in migrating corrosion inhibiting systems.

Figure 10.3 shows some of their results, though it should be noted that after 12 months the potential values for the treated system, while lower, were approaching –350 mV (equivalent to –450 mV approx Cu/CuSO4). There are a number of reasons why measurement of potential may not be a good guide to corrosion behaviour in concrete, including possible poisoning of the reference electrode, if an embedded system is used. For this reason corrosion current measurements are usually more reliable. Figure 10.4 shows the effect on macrocell current, which is a better guide to behaviour, but again does not show anything near the tenfold reduction in corrosion current that would indicate a significantly effective inhibitor. Such values, while confirming some effect, cannot be taken as proof of effectiveness in the field.

In 2003/4 Grantham and co-workers carried out an evaluation of two corrosion inhibitors, Flexcrete Cemprotec MCI 2020 and Sika FerroGard 903, for possible remediation of chloride contamination to a car park

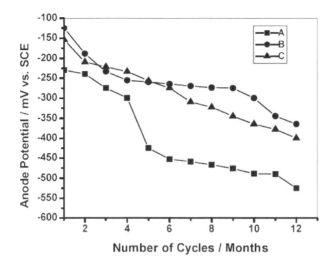

Figure 10.3 Potential vs number of cycles of exposure for steel in OPC concrete under macrocell conditions [A – Control; B – Migrating; C – Admixed] (from Thangavel et al., 2009).

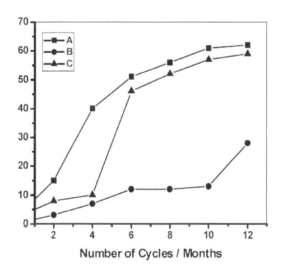

Figure 10.4 Macrocell current vs number of cycles of exposure for steel in OPC concrete under macrocell condition [A – Control; B – Migrating; C – Admixed] (from Thangavel et al., 2009).

deck (Grantham et al., 2005). Tests were performed by coating areas of the concrete with inhibitor and then using a combination of rest potential and galvanostatic pulse monitoring to evaluate their effectiveness. Neither product proved effective and subsequent tests by Sika showed penetration of their product to only about 25 mm, which was the approximate depth of the top of the shallowest steel reinforcement.

A report prepared by Sprinkel (2003) for the Virginia Dept of Transport in the USA concluded, 'In summary, this project does not show any benefit from the use of the corrosion inhibiting admixtures and the topical applications made to the chloride-contaminated concrete surfaces prior to placement of the patches and overlays.' Although it did go on to state that, 'Additional years of monitoring of the exposure slabs and bridges may provide useful results.'

The UK Highways Agency, in its Design Manual for Roads and Bridges, Part 8, BA57/01 (Highways Agency, 2001) refers to research carried out by the Transport Research Laboratory and concludes:

> The TRL research, which was conducted with reasonably good quality concrete, indicates positive results for the effectiveness of inhibitors in the form of cast-in concrete admixtures based on calcium nitrite and amino alcohols, used in new construction. The results for the migrating surface applied and the pelleted delivery system corrosion inhibitors tested is less encouraging. However other researchers have found in tests conducted in lower quality concrete that there may be some beneficial effects with these migrating inhibitors. They may be considered for use when applied to concrete of poor quality, where the chloride levels are low. However for the present their use is not advocated on high quality relatively impermeable structural concrete.

Sika Ltd, manufacturers of Sika Ferroguard 901 and 903 were asked for any data and reports that they had regarding the efficacy of their products for corrosion inhibition. Reviewing the data provided confirmed the view that admixed Ferroguard inhibitor was successful in inhibiting corrosion behaviour, as has been confirmed by other workers. However, the data regarding the behaviour of the 903 penetrating inhibitor product, while confirming that penetration could be achieved, was not sufficiently robust in proving a sufficient effect on corrosion current or potential when used in penetrating mode. However, see also later comments on the possibility of using a pumping method to introduce the inhibitor, as suggested by Cailleux et al. (2008). Sawada and co-workers (Sawada et al., 2005) also attempted to improve the efficiency of injection of ethanolamine and guanidine inhibitors by electrochemical methods, and found that in carbonated concrete, significant improvement in penetration was achieved. In uncarbonated concrete, some improvement was found with guanidine, but little effect was observed with the ethanolamine inhibitor.

Overall, in reviewing the literature and the experience of others in the field, it is apparent that migrating corrosion inhibitors have only limited usefulness in repair while flood coating, brushing or spraying methods are used. They may be successful to some extent in permeable concrete with low cover (say up to 15 mm) which does not contain excessive levels of chloride (and certainly not more than 1% by mass of cement). For other concretes which are less permeable or where the steel is at deeper cover, they are unlikely to penetrate to a sufficient depth to be useful. Indeed, it has been suggested that if they only partially reached the steel reinforcement, they might exacerbate the problem by causing an electrochemical cell to be set up between the area reached by the inhibitor and surrounding anodic areas.

Broomfield (2007), in his book on corrosion of steel in concrete, made the following conclusions regarding penetrating corrosion inhibitors:

- There is very little field data on corrosion inhibitors.
- The available field data is often poor, with no clear evidence of the amount of inhibitor applied, whether it reached the rebar and if it is reducing corrosion rates and extending time to cracking.
- Many claims have been made about the transport of inhibitors through hardened concrete; these need to be independently assessed.
- We will need definitive evidence of the dosage vs. chloride level to achieve a given (low) corrosion rate.
- For application to hardened concrete, we need quantitative data on its penetration vs. concrete cover and concrete permeability.
- We need more information on the performance of inhibitors, particularly well-controlled field trials and long-term corrosion monitoring.

There is little that this author would add to these conclusions. There are better methods for dealing with repair of, especially, chloride-contaminated structures, such as chloride removal, the use of galvanic anodes or a full, impressed current, cathodic protection system and the reader should consider their use in the first instance.

However, the ultrasonic pump idea for introducing inhibitor deeper into a concrete or grout, trialled by Cailleux and co-workers definitely had promise and deserves further investigation.

Acknowledgements

The author would like to acknowledge the help of Dr John Broomfield in preparing this chapter.

References

Broomfield, J. P., *Corrosion of Steel in Concrete – Understanding, Investigation and Repair*, 2nd Edition, London: Taylor and Francis, 2007.

Cailleux, E. 'Treatment of precorroded steel reinforcement by surface-applied corrosion inhibitors: solution tests and application to concrete samples.' *Proceedings of Concrete Solutions, 2nd International Conference on Concrete Repair*, St. Malo, France, BRE Press, 2006.

Cailleux, E., Pollet, V., Dubois, P. M. and Michaux, D. 'A new corrosion treatment for prestressed rebars: the direct injection of a corrosion inhibitor by an ultrasonic pump.' *Proceedings of Structural Faults and Repair*, Edinburgh, 2008.

Grantham, M., Gray, M. and Schneck, U., 'Evaluation of corrosion inhibitors for remediation of St. Mary's multi-storey car park Colchester,' in *Repair and Renovation of Concrete Structures*, Professor Ravindra K. Dhir; Dr M. Rod Jones; Dr Li Zheng (Editors), July 2005.

Highways Agency, *Design Manual for Roads and Bridges*, Part 8, BA57/01, Highways Agency, UK, 2001.

Page, C.L., Ngala, V.T. and Page, M.M., 'Corrosion inhibitors in concrete repair systems,' *Magazine of Concrete Research*, 2000, 52, No. 1, Feb., 25–37.

Raharinaivo, A. and Malric, B., 'Performance Of Monofluorophosphate For Inhibiting Corrosion Of Steel In Reinforced Concrete Structures.' *Proceedings of International Conference on Corrosion and Rehabilitation of Reinforced Concrete Structures*, Orlando, Florida, December 1998.

Sawada, S., Page, C.L. and Page, M.M., 'Electrochemical injection of organic corrosion inhibitors into concrete.' School of Civil Engineering, University of Leeds, Corrosion Science, Elsevier, 2005.

Schiegg, Y., Hunkeler, F. and Ungright, H., 'The effectiveness of corrosion inhibitors, a field study.' *Proceedings of 16th Congress of IABSE*, Lucerne, 2000.

Sprinkel, M., *Final Report – Evaluation of Corrosion Inhibitors for Concrete Bridge Deck Patches and Overlays*. Report VTRC 03-R14, Virginia Transportation Research Council, June 2003.

Thangavel, K., Muralidharan, S., Saraswathy, V., Quraishi, M. A. and Ki Yong Ann. 'Migrating vs admixed corrosion inhibitors for steel in Portland, Pozzolona and slag cement concretes under macro cell conditions,' *Arabian Journal for Science and Engineering*, Volume 34, Number 2C, December 2009.

Vermani, Y. P and Clemena G. G, *Corrosion Protection: Concrete Bridges*, Federal Highways Administration, Virginia, VA: McLean 1998.

11 European standards for concrete repair

Peter Robery

11.1 Introduction

The Single European Act of 1985 started the process of forming a European Single Market, intended to remove barriers to free trade across Europe. Prior to this, each country had its own national standards and there was an understandable level of reluctance to accept products and systems tested to the standards of another country.

This applied equally to the field of concrete repair, which was a relatively new area of concrete technology and one with a bewildering array of materials used singly or in combination to effect repair and protection. While product datasheets for paints and mortars could be translated from one country to another, the test standards and other ad-hoc tests were often not obviously equivalent in the different countries. A simple example is compressive strength, which in the UK is by crushing a cube specimen, whereas on the Continent, cylinders are commonly used, but at least there are established correlations between the two. For more complex properties, such as the carbon dioxide diffusion resistance of a paint film, methods of test and interpretation varied widely, coupled with many practitioners not understanding the significance of many of the test results that appeared on product data sheets.

The situation improved with the development of test procedures by the European Organisation for Technical Approvals (EOTA). In the UK, the British Board of Agrément developed testing procedures for repair mortars and then whole repair systems, which gave the specifier some confidence that an independent party had checked, verified and tested the products to confirm they were following best practice. However, these Certificates were also national, and were based on the national standards and test methods of the originating EOTA organisation. Manufacturers found they often needed EOTA certificates issued for each country in which they were operating.

The scope of the Single European Act included the development of a series of 'New Approach' Directives. These were to provide a framework for a single market in goods and services, based on a series of supporting technical specifications, or standards. The Construction Products Directives 17 and

18 (European Parliament, 2004) addressed the area of concrete protection and repair and in response CEN, the European standards body, set up Subcommittee TC104/SC8 in 1988 to develop the necessary performance specifications and test methods for the industry.

The concrete repair sector in Europe was relatively new in 1988 and few standards applied directly to concrete repair products and systems. The concrete repair sector would therefore provide an ideal route for standards development. Representatives from CEN member countries across Europe nominated technical representatives that could support Steering Committees and Working Groups tasked with drafting the performance standards and test methods that were needed.

European Standard EN 1504 'Products and systems for the protection and repair of concrete structures' is the result of this labour, which came into effect on 31 December 2008, taking 20 years to complete. The standard comprises 10 Parts (Table 11.1), six of which are performance standards (Parts 2 to 7) that set minimum performance criteria for repair products and systems to be used for protection and repair of both reinforced and unreinforced concrete.

The performance standards cross-reference to nearly 100 new specially developed testing standards for repair products and systems, as well as adopting appropriate existing ISO or national standards (Davies and Robery, 2006). EN 1504 also sets out a methodology for the successful repair and protection of concrete structures, including the supervision and quality control of site works. The end result is a range of products and systems for the protection and repair of concrete that are approved as meeting minimum requirements as 'fit for purpose'. The vehicle for product approval is the 'CE-mark', demonstrating that the product or system meets a particular set of performance requirements.

Compliance with EN 1504 will, in effect, be mandatory for many specifiers and purchasers because of the effect of the European Public Purchasing Directives. Government bodies across Europe, including Highways Authorities, have contributed to the development of EN 1504 over the years and have modified their own specifications to bring them into line with the methodology. EN 1504 is therefore highly relevant to the owner, designer, specifier and contractor as well as to the materials supplier.

The EN 1504 series and associated test methods were drafted by CEN committee TC104/SC8 and the standards apply across the 25 member countries of the EU as well as Switzerland, Norway and Iceland.

11.2 Structure of the Standard

EN 1504 builds on early work on concrete repair published previously in Europe, including the RILEM Technical Committee on repair strategy (Schiessl, 1994), German guidelines on repair (DAfSTb, 1991), reports by the UK's Building Research Establishment (BRE, 1997) and various Concrete Society Reports (Concrete Society, various dates), among others. EN 1504 comprises ten Parts, as set out in Table 11.1, with the Parts splitting into the following five sub-groups:

- Part 1 provides a series of definitions.
- Parts 2 to 7 are the product performance specifications for a variety of repair and maintenance tasks, including protective coatings, repair mortars, plate bonding adhesives and crack injection materials.
- Part 8 addresses manufacturer Quality Control systems.
- Part 9 provides the general methodology of repair, from initial diagnosis, through selection of the most appropriate repair options for the particular circumstances and client needs, to specification of the minimum performance requirements for specific repair products and systems.
- Part 10 covers methodology of site installation for the products and systems, including site testing, supervision and quality control.

Part 9 is the starting-point for the series, as it puts the other parts of EN 1504 into context and cross-references to the other Parts. It was the first Part of EN 1504 to be drafted and was issued in 1997 as ENV1504-9:1997, a 'voluntary standard', inviting public comment before it was finalised, and the approach was justified on three grounds:

- The design approach was the key to the success of the whole repair process and should be evaluated in industry before issuing in a mandatory form.
- Other parts of the standard had yet to be drafted and completed, and so the final scope and content of Part 9 could be changed by the drafting teams.
- In any case, without other parts being published, product approval and CE-marking of products and systems could not begin, making compliance with Part 9 (relating to use of 'CE-marked' products) an impossible task.

Table 11.1 Full titles of EN 1504 Parts 1 to 10

Part 1	General scope and definitions, published 2005
Part 2	Surface protection systems, published 2004
Part 3	Structural and non-structural repair, published 2005
Part 4	Structural bonding, published 2004
Part 5	Concrete injection, published 2004
Part 6	Grouting to anchor reinforcement or to fill external voids, published 2007
Part 7	Reinforcement corrosion prevention, published 2007
Part 8	Quality control and evaluation of conformity, published 2004
Part 9	General principles for the use of products and systems, published 2008
Part 10	Application of products & systems and quality control of the works, published 2003

As the years passed, other Parts of the Standard were finalised by the CEN internal enquiry and approval process and then issued, such that by 2006 ENV1504-9 needed to be revised by TC104/SC8 to bring it into alignment with the latest thinking on the performance requirements for products and systems for protection and repair. Over the period from 1990, several significant advances had also been made in the technology of repair and strengthening concrete structures, including the use of carbon fibre plates, strips and wraps for structural strengthening of members, development of topically applied corrosion inhibitors and the increasingly widespread use of impressed current cathodic protection systems.

In parallel with development of the performance standards, TC104/SC8 and its working groups also prepared standards for the supporting test methods, which would be used to evaluate the performance of the products and systems. The purpose was to select a single method of test for each performance requirement. CEN has a strict procedure that its working groups must follow when considering whether new draft standards need to be developed, as set out below:

- first, whether an existing ISO test method can be used;
- second, whether an existing CEN member country's national standard can be used;
- third, whether another method can be developed from a recognised body, such as RILEM, DAfSTb or BRE or from other research sources.

In a few cases, well-established ISO or national methods could be selected, either used as they were, or slightly adapted; examples include the epoxide equivalent of resin adhesives and the viscosity of paints.

In most cases, the existing test methods needed significant adaptation for use with the formulations of repair products and systems on the market. For example, EN 12190 *Determination of compressive strength of repair mortar* is based on BS 6319 *Testing of resin and polymer/cement compositions for use in construction: Part 2 Compressive Strength*. Other test standards have been re-written from national or learned guidance, such as the RILEM Thermal Compatibility Tests I & II for paints applied to concrete, which have been adapted for use on overlays of concrete repair mortar bonded onto a reference substrate, creating EN 13687 *Determination of thermal compatibility*; EN 13687-1 is based on the RILEM freeze–thaw cycling test with de-icing salt immersion and EN 13687-2 is the thunder shower (thermal shock) test.

In other cases, TC104/SC8 agreed that no appropriate test method existed for measuring the specialist properties of repair products or systems. New methods of test were drafted based on the best available practice around the world and the drafts circulated among CEN member countries until universal agreement was achieved, often by pooling research knowledge; examples include EN 13396, *Measurement of chloride ion ingress (bulk diffusion method)*, EN 1062-6, *Paints and varnishes – Coating materials and coating systems for exterior masonry – Part 6: Determination of carbon dioxide permeability*, and EN 13395-3, *Determination of workability – Part 3: Test for flow of repair concrete*, which is based on the UK's Highways Agency specification. A selection of the test standards is given in the references, with a full list contained in each Part of EN 1504 and in the paper by Davies (Davies and Robery, 2006).

Revision to Part 9 began in 2006, with the final version issued for CEN approval in October 2008, so completing the revision task. EN 1504 came into force on 31 December 2008, on which date all conflicting national performance standards and test methods across Europe were withdrawn.

The package of performance and test standards covers the whole repair enterprise, enabling specifiers, engineers, contractors and others to make genuine comparisons of performance based on common data, linked to a series of performance specifications for the products.

11.3 EN 1504-9 – General Principles for Protection and Repair

From the outset, TC104/SC8 agreed that EN 1504 should not be a code of practice but a methodology for the repair process; anything more could put the standard in conflict with various national regulations of the CEN member countries. The end result is designed to give guidance on the causes of deterioration and the need for repair and deliver a repair package using products and systems tested to European standard methods and approved by the 'CE-mark', demonstrating that the product or system meets a particular set of performance requirements. The standard cannot identify the role of the structure within the client's own strategy or business, or external constraints such as planning rules, environmental limits on construction works or conflicts between the client's business processes and certain repair options. Nor can the standard identify the value of the building to the client. These are within the remit of the engineer and asset manager and will vary nationally. Also, the standard is not intended to be a repair manual and for example, does not give details on how a structure should be inspected, tested, diagnosed and structurally appraised. Other documents give appropriate levels of guidance (e.g. Broomfield, 2007, Bungey et al., 2006, Concrete Society TR 54, 2000).

Regarding methodology, EN 1504 has two parts reflecting design and execution: EN 1504-9 *General principles for the use of products and systems*; and EN 1504-10 *Site application of products and systems and quality control of works*.

EN 1504-9 was conceived to provide:

* a logical repair methodology
* a technical standard structured so as to allow the client to exercise economic choices based upon whole-life costing

- a framework for the specification of repair products covered by other parts of this and other relevant standards.

Table 11.2 lists the main clause headings in EN 1504-9, as issued on 31 December 2008. From Table 11.2 it can be seen that EN 1504-9 sets out the following methodology:

- assessment of the structure and the extent of damage;
- determination of the objectives of the repair work;
- definition of the available principles for dealing with the identified deterioration in accordance with the stated objectives;
- identification of the specific properties required of the materials to be used in the repair.

Table 11.3 summarises the common steps in the process of assessment, specification, site execution and maintenance and monitoring of structures for repair, as set out in EN 1504-9. Clause numbers to the relevant texts in EN 1504-9, along with other Parts of the EN 1504 series as appropriate, are also listed in column 1.

EN 1504-9 is intended to be sufficiently flexible to be used in the various regulatory and contractual environments within Europe. It is drafted with a normative text, supplemented by an informative Annex A that has common clause numbers with the normative section and provides informative comment on the normative text. For those not familiar with the language of European Standards, an informative Annex gives supplementary information, the use

Table 11.2 Clause numbers and titles in ENV 1504-9

Clause number	Title
1	Scope
2	Normative references
3	Definitions
4	Minimum requirements before protection and repair
5	Protection and repair within a Structure Management Strategy
6	Basis for the choice of protection and repair principles and methods
7	Properties of products and systems required for compliance with the principles of protection and repair
8	Maintenance following completion of protection and repair
9	Health, safety and the environment
10	Competence of personnel
Annex A	Informative guidance and background information

Table 11.3 Six steps in assessment, specification, repair and maintenance

Step 1: Information about the structure Cl 4	• Present condition and history of the structure • Original design approach • Available documentation • Performance of previous repair and maintenance work • Environment, loading and contamination during construction and use • Requirements for future use
Step 2: Process of assessment Cl 4	• Assessment of defects and their causes (Figure 11.1) • Safety and structural appraisal in the current condition • Cost and funding of alternative protection or repair options, including future maintenance and access costs • Consider the appearance of the protected or repaired structure
Step 3: Management strategy Cl 5 & 6	• Undertake a safety and structural appraisal for the structure for the cases of before, during and after protection and repair work • Identify options for each repair case (Table 11.4) • Define principles for repair and protection to combat the deterioration mechanisms, acting singly or in combination (Table 11.5) • Select the methods for repair and strengthening (Table 11.5) • Select the methods of protection to combat the exposure environment
Step 4: Design of repair work Cl 6, 7 & 9 EN 1504: Pt 2–7	• Consider the intended use of products and systems • Consider ease of maintenance and re-repair, if required • Assess application requirements for substrate, products and working methods • Prepare specifications and work drawings • Undertake a safety and structural reappraisal after completion of the protection and repair work
Step 5: Repair work Cl 6, 7, 9 & 10 EN 1504: Pt 10	• Selection and use of materials and equipment in the prevailing work environment, including minimum performance levels • Quality Control tests • Health and Safety of stakeholders • Competence of personnel, including trial repairs
Step 6 Acceptance Cl 8 EN 1504: Pt 10	• Product sampling and identification testing • Acceptance testing following application • Remedial action required to the products or systems • Future maintenance schedule • Full works documentation

Table 11.4 Six options for tackling a deteriorated concrete structure

•	Do nothing for a certain time but monitor
•	Re-analyse the structural capacity, possibly leading to downgrading in function
•	Prevent or reduce further deterioration
•	Strengthen or repair and protect all or part of the concrete structure
•	Reconstruct or replace all or part of the concrete structure
•	Demolish all or part of the concrete structure

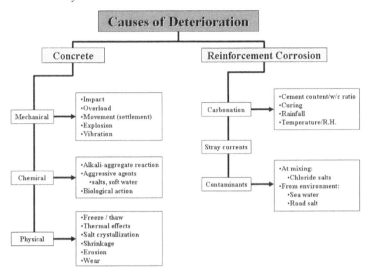

Figure 11.1 Common causes of concrete deterioration covered by EN 1504 methods

of which is not mandatory, but it may assist with the interpretation of the normative text.

11.4 The options, principles and methods of repair

The inter-relation between EN 1504-9 and other parts of the EN 1504 series is set out in Clause 6, which contains a large table giving various repair principles and corresponding methods to achieve the repair principles. Table 11.5 is reproduced from Clause 6 of EN 1504-9 and is split into two main groupings, namely:

- Table 11.5a: repair and protection to damaged concrete;
- Table 11.5b: repair and protection to concrete damaged by corrosion of embedded reinforcing steel.

Table 11.5 lists 11 repair principles and 43 repair methods offering different ways of achieving a given principle. A requirement placed on TC104/SC8, through the Construction Products Directive, was to ensure the list of methods reflects all products that are currently offered for sale by manufacturers in Europe. No existing product that was on the market should be excluded by the new EN 1504 Standard. Some of the methods (e.g. Method 7.4) are less familiar than others.

The third column in Table 11.5 generally lists the relevant Part of EN 1504 where the minimum performance requirements and other properties

Table 11.5a Repair principles and methods based on EN 1504-9

Principle	Examples of methods based on the principles	Relevant standard or other route
Principles and methods related to defects in concrete		
1. Protection against ingress	1.1 Hydrophobic impregnation	EN 1504-2
	1.2 Impregnation	EN 1504-2
	1.3 Coating	EN 1504-2
	1.4 Surface bandaging of cracks	EOTA
	1.5 Filling of cracks	EN 1504-5
	1.6 Transferring cracks into joints	EOTA
	1.7 Erecting external panels	EN1992-1-1
	1.8 Applying membranes	EOTA
2. Moisture control	2.1 Hydrophobic impregnation	EN 1504-2
	2.2 Impregnation	EN 1504-2
	2.3 Coating	EN 1504-2
	2.4 Erecting external panels	EN1992-1-1
	2.5 Electrochemical treatment	Proprietary
3. Concrete restoration	3.1 Hand-applied mortar	EN 1504-3
	3.2 Recasting with concrete or mortar	EN 1504-3
	3.3 Spraying concrete or mortar	EN 1504-3
	3.4 Replacing elements	EN1992-1-1
4. Structural strengthening	4.1 Adding or replacing embedded or external reinforcing bars	EN1992-1-1
	4.2 Adding reinforcement anchored in pre-formed or drilled holes	EN 1504-6
	4.3 Bonding plate reinforcement	EN 1504-4
	4.4 Adding mortar or concrete	EN 1504-3 & -4
	4.5 Injecting cracks, voids or interstices	EN 1504-5
	4.6 Filling cracks, voids or interstices	EN 1504-5
	4.7 Prestressing – (post-tensioning)	EN1992-1-1
5. Increasing physical resistance	5.1 Coating	EN 1504-2
	5.2 Impregnation	EN 1504-2
	5.3 Adding mortar or concrete	EN 1504-3
6. Resistance to chemicals	6.1 Coating	EN 1504-2
	6.2 Impregnation	EN 1504-2
	6.3 Adding mortar or concrete	EN 1504-3

Table 11.5b Principles and methods related to reinforcement corrosion

7. Preserving or restoring passivity	7.1 Increasing cover with additional mortar or concrete	EN 1504-3
	7.2 Replacing contaminated or carbonated concrete	EN 1504-3
	7.3 Electrochemical realkalisation of carbonated concrete	CEN/TS 14038-1
	7.4 Realkalisation of carbonated concrete by diffusion	Proprietary
	7.5 Electrochemical chloride extraction	CEN/TS 14038-2 (draft)
8. Increasing resistivity	8.1 Hydrophobic impregnation	EN 1504-2
	8.2 Impregnation	EN 1504-2
	8.3 Coating	EN 1504-2
9. Cathodic control	9.1 Limiting oxygen content (at the cathode) by saturation or surface coating	Proprietary
10. Cathodic protection	10.1 Applying an electrical potential	EN12696-1
11. Control of anodic areas	11.1 Active coating of the reinforcement	EN 1504-7
	11.2 Barrier coating of the reinforcement	EN 1504-7
	11.3 Applying corrosion inhibitors in or to the concrete	EOTA

can be found for products and systems for any given method. For example, the requirements for Method 3.1, hand-applied mortar, can be found in EN 1504-3 (i.e. Part 3). In EN 1504-9, the third column is left blank for methods that are outside EN 1504, but Table 11.5 has been completed here with three additional categories: other relevant *European Standards;* EOTA-certified techniques that are outside EN 1504; and Proprietary systems offered by one or more companies with no independent performance certification at this time.

The following examples give more information on the application of some of these alternative methods:

- *1.4 Surface bandaging of cracks* – waterproofing systems for cracked concrete comprising glued rubber sheeting bonded onto concrete surfaces are currently outside EN 1504, but are covered by EOTA certification.
- *1.7 Erecting external panels* – this can be achieved by designing a new precast or in-situ concrete wall and the design of such a wall is covered by an existing Eurocode Design Standard EN 1992-1-1.
- *2.5 Electrochemical treatment* – electrochemical drying of concrete by passing a current through the reinforcement, similar in principle

to cathodic protection; currently only proprietary systems are available.

- *3.3 Spraying concrete or mortar* – while the performance of the finished product is covered in EN 1504-3, another Standard specifically deals with sprayed concrete generally: EN 14487-1.
- *7.3 Electrochemical realkalisation of carbonated concrete* – the requirements are set out in a new performance standard specifically for realkalisation and chloride extraction in reinforced concrete in CEN/TS 14038-1.
- *10.1 Applying an electrical potential* – the requirements are set out in a new performance standard specifically for cathodic protection of reinforced concrete structures in EN 12696.

Once the problem is defined and the extent of the damage known, the designer and specifier can engage in an informed discussion about the possible options for repair and the potential costs and appropriate timescales. Table 11.4 lists the six common options for repair found in EN 1504-9. An effective repair strategy will normally incorporate different options for repair based on a range of factors such as the orientation of the structure (e.g. to the sea), its height above ground, variation in the quality of concrete around the structure, the variation in the cover to the reinforcement and many more.

The standard does not and could not give definitive guidance on the durability of the various options, or the relative performance of the various products available to meet the options, principles and methods, beyond setting the minimum performance criteria.

11.5 Performance Standards (parts 2–7)

As noted above, while EN 1504-9 gives 'General Principles' for repair, the main purpose of EN 1504 is as a product standard leading to 'CE-marking' of products and systems that are suitable for the protection and repair of concrete structures.

To achieve a CE-mark, products and systems have to reach minimum performance standards for a range of engineering properties, related to the method to be used (Table 11.5). For example, a surface protection system for concrete, such as a film-forming paint, will have different performance requirements depending on whether it is intended to protect against ingress of chloride ions, reduce carbonation of the concrete, control moisture penetration, or whether the paint is to be applied over active cracks in the concrete.

CE-marking is intended to deliver products and systems certified to meet minimum criteria for one (or more) of the repair methods listed in Table 11.5 (e.g. a very high performance coating may meet or exceed the minimum performance for several methods, yet a lower performance product may only be CE-marked for one method).

CE-marking also ensures that the products and systems sold are certified as safe (i.e. in terms of release of dangerous substances and reaction to fire) and consistent (i.e. certified under a factory production control system to deliver a quality assured product). This is set out in EN 1504-8, covering quality control and evaluation of conformity for the products and systems.

The most important aspect for the specifier or end user is the performance of the product or system in service. The specifier must be aware that the performance standards give a minimum level of performance, below which the product will not be 'fit for purpose' and therefore will not carry a CE-mark for the repair method. While this minimum performance may be suitable for a 'general' intended purpose, it may not necessarily be adequate for all applications.

When the appropriate repair methods have been selected, the relevant performance requirements can then be built into the specification for the works. Traditionally, full and detailed performance specifications have been needed to prevent unsuitable products being used. EN 1504 overcomes this, as the CE-marking process ensures that only suitable repair products and systems are available to satisfy the various methods. The specifier need only quote in the specification that a product should be CE-marked for the required method (e.g. Method 7.2 *Replacing contaminated or carbonated concrete*).

In reality, the EN 1504 specification process is slightly more complicated. Parts 2 to 7 of EN 1504, which contain the performance requirements, prescribe minimum performance acceptance values for specific critical properties of the products and systems. Each of the Parts lists three classifications of performance requirement:

• All intended uses – minimum performance values that every product must meet for the particular method, including physical and chemical properties, quality control and environmental issues;
• Certain intended uses – recommended minimum performance values for one or more additional tests that may be called up by the specifier for an unusual or difficult repair situation;
• Special applications – recommended methods of test, with no recommended minimum performance values, that may be called up by the specifier for a very unusual or difficult repair situation and are found in the informative Annex B to the performance standards.

Use of EN 1504 in a specification is best illustrated by an example, such as for Method 7.2 above. Table 11.6 is an extract from EN 1504-3, listing the performance requirements for four strength classes of mortar (Classes R1 to R4) and corresponding performance requirements for all intended uses.

Depending on the particular project, the engineer may decide that the repair mortar should be stronger than a class R4 mix in Table 11.6, to match the strength of the existing concrete. The specification for Method 7.2 could

Table 11.6 Method 7.2, minimum requirements for all intended uses (based on EN 1504-3)

Item	Performance characteristic	Test method	Requirement			
			Structural		Non-structural	
			Class R4	Class R3	Class R2	Class R1
1	Compressive Strength	EN 12190	≥45 MPa	≥25 MPa	≥15 MPa	≥10 MPa
2	Chloride content	EN 1015-17	≤ 0.05%		≤ 0.05%	
3	Adhesive bond*	EN 1542	≥2.0 MPa	≥1.5 MPa	≥0.8MPa	
4	Restrained* shrinkage	EN 12617-4	Bond strength after test			Not required
			≥2.0 MPa	≥1.5 MPa	≥0.8 MPa	
5	Carbonation resistance	EN 13295	dk ≤ control concrete		Not required	

* Product or system applied to a reference concrete substrate to EN 1766

be for a CE-marked product that met the requirements of Method 7.2, but with a minimum strength of 60 MPa.

Table 11.7 can be used to add additional properties for the repair mortar, for example where the specific requirements of the project mean that a successful repair application is more demanding. These may be required in various situations, such as the repair having a structural requirement (elastic modulus), which may be important because some resin mortar products may have a modulus that is significantly lower than 20 GPa for Class R4. Also, if the repair is in an area with particularly severe thermal cycles, then Items 7, 8 or 9 could be added, reflecting a severe northern European climate, a wet Mediterranean climate, or a dry southern Mediterranean climate respectively.

It should be remembered that the performance standards do not limit the materials that can be used in the products or systems. In the case of repair mortar, the products are classified into three types (EN 1504-3):

- CC – mortars made using only hydraulic cements as a binder, as defined in EN197
- PCC – mortars comprising hydraulic cements and polymer dispersions, also-called 'polymer-modified cementitious mortar'
- PC – mortars made using polymer binders only, such as epoxy, acrylic or polyester resin.

Now that the CE-marking process has started, new products have begun to appear on the market carrying the specified label and CE-mark, as set out in Annex ZA of each performance standard. An example of a typical label is shown in Figure 11.2. Many manufacturers have opted to test their products and systems for both sets of requirements in the relevant performance

Table 11.7 Method 7.2, minimum requirements for certain intended uses (EN 1504-3)

Item	Performance characteristic	Test method	Requirement			
			Structural		Non-structural	
			Class R4	Class R3	Class R2	Class R1
6	Elastic modulus	EN 13412	≥20 GPa	≥15 GPa	No requirement	
7	Thermal compatibility Pt 1, Freeze–thaw*	EN 13687-1	Bond strength after 50 cycles			Visual inspection after 50 cycles
			≥2.0 MPa	≥1.5 MPa	≥0.8 MPa	
8	Thermal compatibility Pt 2, Thunder shower*	EN 13687-2	Bond strength after 50 cycles			
			≥2.0 MPa	≥1.5 MPa	≥0.8 MPa	
9	Thermal compatibility Pt 4, Dry cycling*	EN 13687-4	Bond strength after 50 cycles			
			≥2.0 MPa	≥1.5 MPa	≥0.8 MPa	
10	Skid resistance*	EN 13036-4	Class I: >40 units wet tested Class II: >40 units dry tested Class III: >55 units wet tested			
11	Coefficient of thermal expansion	EN 1770	Not required if tests 7, 8 or 9 are carried out, otherwise declared value			
12	Capillary absorption	EN 13057	≤ 0.5 kg m^{-2} hour$^{-0.5}$			No requirement

* product or system applied to a reference concrete substrate to EN 1766

standards (i.e. for all intended uses and certain intended uses, as listed in Tables 11.6 and 11.7).

It is a requirement of the CE-marking process that the label only carries the fact that the product passes the test; the specifier would have to refer to the manufacturer's data sheet to find the actual test results for the individual test method.

11.6 Summary

Repair and rehabilitation plays an important part in the effective management of concrete assets, but opinions have differed as to the best way to tackle a deteriorating concrete structure. Repair practice is rapidly changing, with a constant stream of new products and systems available, bringing a bewildering range of claims before the architect, designer and specifier.

The new European Standard EN 1504 'Products and systems for the protection and repair of concrete structures', has been developed to help the specifier choose the correct repair products and systems for the job and the requirements of the Construction Products Directive, as realised through the

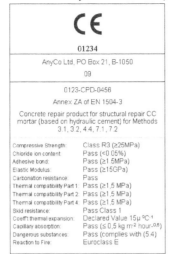

$C\mathcal{E}$

01234

AnyCo Ltd, PO Box 21, B-1050

09

0123-CPD-0456

Annex ZA of EN 1504-3

Concrete repair product for structural repair CC
mortar (based on hydraulic cement) for Methods
3.1, 3.2, 4.4, 7.1, 7.2

Compressive Strength:	Class R3 (≥25MPa)
Chloride ion content:	Pass (<0.05%)
Adhesive bond:	Pass (≥1.5MPa)
Elastic Modulus:	Pass (≥15GPa)
Carbonation resistance:	Pass
Thermal compatibility Part 1:	Pass (≥1.5 MPa)
Thermal compatibility Part 2:	Pass (≥1.5 MPa)
Thermal compatibility Part 4:	Pass (≥1.5 MPa)
Skid resistance:	Pass Class 1
Coeff't thermal expansion:	Declared Value 15μ °C⁻¹
Capillary absorption:	Pass (≤ 0.5 kg m⁻² hour⁻⁰·⁵)
Dangerous substances:	Pass (complies with (5.4)
Reaction to Fire:	Euroclass E

Figure 11.2 A typical CE-mark label for a product to methods 3.1, 3.2, 4.4, 7.1 and 7.2.

CE-marking to the EN 1504 series, has begun. This process is set to change the future direction of the concrete repair industry.

EN 1504 will affect the asset owner, as well as the specifier, contractor and manufacturer. For the first time a Standard exists that addresses all stages of the repair process, from initial awareness of a problem, to the handover of a properly designed and executed repair of the building to a satisfied client, incorporating the use of products and systems that are approved as meeting minimum performance requirements for a range of repair applications. However, guidance is still needed on its use and interpretation. EN 1504 assumes the reader is competent in the field of concrete inspection, diagnosis, repair scheme development and material specification, which may not necessarily be the case.

Further details of EN 1504 may be obtained from the UK Concrete Repair Association website and joint CPA/Institute of Corrosion/Concrete Society publications.

The views expressed in the chapters of this book are those of the individual authors. While every effort has been made to ensure that the information therein is up to date and factually correct, the Publisher and Editor do not accept any responsibility for the contents of the chapters or for any loss or damage that might occur as a result of using the data or advice given in the chapters.

Permission to reproduce extracts from BS EN 1504 is granted by BSI. British Standards can be obtained in PDF or hard copy formats from the BSI online shop: www.bsigroup.com/Shop or by contacting BSI Customer Services for hardcopies only: Tel: +44 (0)20 8996 9001, Email: cservices@bsigroup.com.

Bibliography

BRE, *Progress in European Standardisation for the Protection and Repair of Concrete*, Information Paper IP11/97, 1997.

BRE Digest 444, Parts 1–4, *Corrosion of Steel in Concrete*, 2000.

Broomfield, J. P., *Corrosion of Steel in concrete* 2nd Edn, Taylor and Francis, 2007.

Bungey, J. H., Millard, S. G. and Grantham, M. G., *Testing of Concrete in Structures*, Taylor & Francis, 4th Edition, 2006.

CEN/TS 14038-1, Electrochemical realkalization and chloride extraction treatments for reinforced concrete – Part 1: Realkalization.

CEN/TS 14038-2, Electrochemical realkalization and chloride extraction treatments for reinforced concrete – Part 2: Chloride Extraction (Draft published 2010).

CIRIA, *Corrosion Damaged Concrete – Assessment and Repair*, Butterworths, 1987.

Concrete Bridge Development Group, *Guide to Testing and Monitoring the Durability of Concrete Structures*, CBDG Technical Guide 2, 2002.

Concrete Society, *Cathodic protection of Reinforced Concrete*, TR36, 1990.

Concrete Society, *Guidance on Radar Testing of Concrete Structures*, TR48, 1997.

Concrete Society, *Guide to Surface Treatments for Protection and Enhancement of Concrete*, TR50, 1999.

Concrete Society, *Diagnosis of Deterioration in Concrete Structures*, TR54, 2000.

Concrete Society, *Electrochemical Tests for Reinforcement Corrosion*, TR60, 2004.

Concrete Society, *Assessment, Design and Repair of Fire-damaged Concrete Structures*, TR68, 2008.

Concrete Society, *Repair of Concrete Structures with Reference to BS EN 1504*, TR69, 2009

Concrete Society, *Concrete Petrography*, TR71, 2010.

Concrete Society Technical Reports on Concrete Repair:

  TR26, 'Repair of concrete damaged by reinforcement corrosion'

  TR33, 'Assessment and repair of fire-damaged concrete structures'

  TR36, 'Cathodic protection of reinforced concrete'

  TR38, 'Patch repair of reinforced concrete subject to reinforcement corrosion'

  TR50, 'Guide to surface treatments for protection and enhancement of concrete'.

DAfStb, The German Committee on Reinforced Concrete. *Guidelines for the Protection and Repair of Concrete Components*, Parts 1 to 4, February 1991.

Davies, H. and Robery, P. C., 'European standards for repair and protection of concrete.', *Proc 2nd Intl Conf on Concrete Repair*, St.Malo, Brittany, 27–29 June 2006.

Directive 2004/17/EC of the European Parliament and of the Council of 31 March 2004 coordinating the procurement procedures of entities operating in the water, energy, transport and postal services sectors.

Directive 2004/18/EC of the European Parliament and of the Council of 31 March 2004 on the coordination of procedures for the award of public works contracts, public supply contracts and public service contracts.

EN 197-1, Cement. Composition, specifications and conformity criteria for common cements, 2000.

EN 1504 Series 'Products and systems for the protection and repair of concrete structures – Definitions, requirements, quality control and evaluation of conformity', implemented 31 December 2008.

EN 1992-1-1, Eurocode 2: Design of concrete structures – Part 1: General rules and rules for buildings.

EN 12696: 2000, Cathodic protection of steel in concrete.

EN 14487-1, Sprayed Concrete – Part 1: Definitions, Specifications and Conformity.

ENV 1504: Part 9: 1997, 'Products and systems for the protection and repair of concrete structures – Definitions, requirements, quality control and evaluation of conformity, Part 9: General principles for the use of products and systems', 1997.

Institution of Civil Engineers National Steering Committee *Recommendations for the Inspection, Maintenance and Management of Car Park Structures*, ICE NSC Report, 2005.

Schiessl, P., 'Draft Recommendation for repair strategies for concrete structures damaged by reinforcement corrosion'. *Materials & Structures*, Vol 27, 1994, 415–436.

St John, D.A., Poole, A. W. and Sims, I., *Concrete Petrography; A Handbook Of Investigative Techniques*, London, Arnold, 1998.

Selected Test Standards called up in EN 1504-3

For a complete list see Davies and Robery, 2006

EN 1542, Measurement of bond strength by pull-off

EN 1766, Reference concretes for testing

EN 1770, Determination of coefficient of thermal expansion

EN 12190, Determination of compressive strength of repair mortar

EN 12617-4, Part 4: Determination of shrinkage and expansion

EN 13036-4, Surface characteristics – Test methods – Part 4: Method for measurement of skid resistance of a surface: The pendulum test

EN 13057, Determination of resistance of capillary absorption

EN 13295, Determination of resistance to carbonation

EN 13395-1, Determination of workability – Part 1: Test of flow of thixotropic repair mortars

EN 13395-2, Determination of workability – Part 2: Test for flow of grout or mortar

EN 13395-3, Determination of workability – Part 3: Test for flow of repair concrete

EN 13395-4, Determination of workability – Part 4: Application of repair mortar overhead

EN 13396, Measurement of chloride ion ingress

EN 13412, Determination of modulus of elasticity in compression

EN 13529, Resistance to severe chemical attack

EN 13584, Creep in compression

EN 13687-1, Determination of thermal compatibility – Part 1: Freeze–thaw cycling with de-icing salt immersion

EN 13687-2, Determination of thermal compatibility – Part 2: Thunder shower cycling (thermal shock)

EN 13687-4, Determination of thermal compatibility – Part 4: Dry thermal cycling

EN 14629, Determination of chloride content in hardened concrete

EN 14630, Determination of carbonation depth in hardened concrete by the phenolphthalein method

Note: In the above list of references, the European Standards are issued by national standards bodies in the CEN member states. In the UK, this is the British Standards Institution (BSI) and so all EN standards in the UK are prefixed with the term BS (i.e. BS EN 1504). Voluntary standards are prefixed by BSI with the letters DD (i.e. DD ENV 1504:1997).

12 Concrete repair – a contractor's perspective

John Drewett

12.1 Introduction

The concrete repair industry has come a long way since 1954 when Concrete Repairs Ltd (CRL) was established as a contractor. Until the early 1980s the majority of repairs were to buildings and marine structures. The work consisted of breakout using hand pneumatic tools and reinstatement using site batched repair mortars and concretes with waterproof additives, such as styrene butadiene rubber (SBR) polymers, to enhance durability.

In the 1980s pre-bagged repair materials using freeze-dried polymers started to become available which improved the quality control for the repairs, and complete repair systems were marketed by suppliers. These consisted of steel and concrete primers, repair concretes and mortars, and finish coating systems.

For the first time clients could specify with some confidence a complete repair system from one supplier. But specifiers were still struggling to quantify the works in this growing industry due to lack of experience and no acknowledged method of measurement. It was very common for concrete repair projects to go way over budget due to poorly constructed bills of quantities and underestimates in quantifying the work.

The market for concrete repair in the UK has grown in the last 20 years as the concrete structures of the 1950s and 60s have matured and it was realised that chlorides from de-icing salts and admixtures such as calcium chloride have initiated rapid corrosion of the reinforcement.

There has been a great deal of research into concrete deterioration and in particular corrosion of the reinforcement. This has led to the development of corrosion control techniques and improved repair systems which have brought about a step change in the way we deal with concrete repair and produced harmonised European Standards.

This chapter will look at how concrete repair is quantified, specified, procured and executed but first of all it will address the importance of safety.

12.2 Safety

A successful contract must never compromise on safety for the sake of commercial gain. CRL has been awarded the RoSPA (Royal Society for the Prevention of Accidents) Gold Award for ten consecutive years but this has not been at the expense of profitability.

Safety, or rather 'health and safety' as it is now referred to (illustrating the importance of protecting people's physical and psychological wellbeing), now forms an integral part of the construction process from the earliest planning stage right through to the use, maintenance and end of life of a building or structure.

To help us through this process we have the likes of the Construction (Design and Management) Regulations, coupled with the vast variety of ACoPs (Approved Code of Practice) and industry guidance now available (mostly free).

With all this information the industry is well equipped to fully understand the acceptable health and safety standards, individual roles and responsibilities and expected outcomes.

Clients, designers, principal contractors and contractors (not forgetting the CDM coordinator for the larger projects) are working together on projects and taking their responsibilities seriously, although there is still room for improvement.

The site teams too have developed over the years with developments in safety culture, and are becoming much more aware that their employers will not compromise on health and safety. The site teams are now far better equipped with knowledge, consultation, modern equipment and personal responsibility.

The role of the Safety Advisor is changing. In the past there was much less legislation and the safety 'officer' would simply carry out quick site inspections and write a scathing report. With sites now better trained and managing their own health and safety in the first instance, the Safety Advisor is free to provide support and advice, to carry out on-site testing for the likes of noise, vibration and CoSHH substances and to assist in health surveillance programmes.

The management of health and safety has improved, but still has much to achieve to further reduce accidents and incidents of ill-health in the construction sector.

In the future industry will need to embrace health and safety challenges arising from developments in technology which may include the likes of nanotechnology and emissions from electromagnetic fields.

12.3 Quantifying the work

Projects used to be tendered on a lump-sum basis which relied on the experience of the estimator to value the works. This made it difficult to compare submitted tenders and value additional works. In the 1990s there was a move towards developing bill items for concrete repair using volume measurement in litres but this created its own problems as the repairs were often wrongly quantified.

Today we have the Concrete Repair Association Method of Measurement which is the acknowledged standard for quantifying the concrete repairs. This is a schedule of repair sizes for overhead, wall and floor repairs. The schedule has been used in the UK since 2000 and has proved to be very successful.

The quantification of the work required should come from a desk-top study of the available drawings and the on-site inspection and investigation work. If a close inspection of all the exposed concrete is not possible then a sample section should be used then pro rata for the full area. It is important to add an additional percentage for contingency which may be 10–30%. This additional quantity is based on the experience and confidence of the Surveyor. The CRA Method of Measurement will enable a comparison of tenders and form the basis for the measurement of works during contract execution.

12.4 Specification

The repair specifications for concrete repair are usually based on a particular material supplier who should be a member of the Concrete Repair Association (CRA) with reference to 'or similar approved', or a member of an appropriate similar organisation in other countries.

The recent introduction of the European Standard for concrete repair BS EN 1504, entitled 'Products and systems for the repair and protection of concrete structures', is now only just starting to make an impact on specifications. This standard is mandatory for government-funded projects and will undoubtedly be adopted by the whole industry.

BS EN 1504 covers the whole concrete repair process, from the initial identification of a problem, through to the works on site. It requires that a 'whole-life cost' approach is taken so that the anticipated future needs of the client are met.

The standard consists of 10 parts, each covered by a separate document. It will give the structure owner an increased level of confidence as, for the first time, all issues of concrete repair and protection are addressed by a single integrated European standard.

For further information, the Concrete Society has published a guidance note, Technical Report TR69, to guide consultants and contractors through the application of BS EN 1504. Chapter 11 also details how BS EN 1504 was adopted and how the various parts work.

Material suppliers are providing specifications and products which are compliant with the standard but care must be taken to ensure that the correct specification is selected for the application.

The standard also emphasises the need for a detailed investigation of the concrete prior to specification and the use of corrosion control techniques such as cathodic protection where needed to address durability issues.

12.5 Conditions of contract

The concrete repair industry covers a wide spectrum of market sectors which includes highways, marine, power, utilities, housing, commercial and industrial. Each sector has its preferred conditions of contract so there is no standard for the concrete repair industry.

However, the main ones used in the UK are the JCT, ICE and NEC conditions although some clients will have their own bespoke conditions.

It is important that the correct version of the chosen suite of contracts is used to ensure that it is appropriate for the size and type of works being undertaken. Some contracts will include full or partial design responsibility, and it is important to establish who is undertaking and who is responsible for the design before selecting the contract to be used.

Contracts can be let in several different ways, i.e. lump sum, traditional Bills of Quantities, activity schedules, target cost, cost reimbursable etc, and the client needs to consider at an early stage which is the most suitable contract to use.

It is also desirable to keep to the wording of standard contracts as written so that the emphasis and original intention of the contract is not compromised. It is not unusual to receive a contract that has an amendment document (usually written by lawyers) that is thicker than the original contract, and where the original emphasis and responsibilities of the contract are altered and heavily biased in favour of the Client. These amendments are often not cross-referenced properly and result in contradictions and conflicting clauses.

The standard contracts have generally been drafted so as to clearly lay out the responsibilities of each party and these should remain as written. It is particularly prevalent in Design and Build contracts that the Client will try and pass their responsibilities on to the Contractor!

12.6 Contractor selection

It is extremely important to ensure correct selection of contractors to tender the work. Depending on the client procurement procedures there may be a need to advertise the contract and issue pre-qualification questionnaires to respondents to select a list of suitable contractors.

There is a long list of qualities which can be used to assess the suitability of contractors:

- Financial
- Experience of similar work

- Audited to BS EN 9002 for Quality
- Audited to BS EN 14001 for Environmental
- Audited to BS EN 18001 for Safety
- A member of the CRA or other appropriate body
- Past references
- CSCS-carded operatives (UK only)
- Training
- Safety record
- Professional qualifications of management.

The list can be endless but it is important to select a minimum of three comparable contractors and a maximum of six for larger schemes. Clients often allow a contractor to tender who does not have the experience and qualifications to undertake the work then find that their tender is the lowest and are obliged to award them the contract.

12.7 Tendering the work

With a specification, conditions of contract and an agreed budget it is time to procure the work. Time and timing are important factors.

- The right time for the client for operational reasons. Retail clients will want to avoid Christmas. Educational clients will want the work done during summer closures. Industrial clients will have operational constraints.
- The right time for the contractor. The contractor will not want to be applying specialist coatings during the winter. Marine works should be scheduled to avoid the winter period.
- Sufficient time to tender for the contractor. Usually this will be a minimum of three working weeks and up to six weeks for complex contracts. It is important to remember that the concrete repair contractor will need to seek subcontract prices for access and surface preparation etc. Insufficient time will lead to poor tenders and fewer returned tenders which will be detrimental to both parties.
- Sufficient time for assessment and contract award by the client. The client will need time to assess the tenders and seek clarification from tenderers where necessary. There may be a need for post-tender interviews of the short-listed contractors. The final contract award may be subject to a standstill period of at least ten days to allow for any objections or a consultation period with housing or commercial tenants.

The tender enquiry should be specific on how the tenders will be assessed to ensure that the procurement is fair and legal. Clients have in the past favoured cheapest as best but this has proved to be a false economy which rewards contractors who cut corners on safety, materials, quality and environmental issues.

New procurement procedures have favoured a mix of quality and price to select the preferred contractor. This is often a 60:40 split quality: price or sometimes 70:30. There should be a clear marking system for tender evaluation so the contractor understands what is important to the client. In some cases the tender will be in two stages with contractors failing to meet the required quality standard being eliminated before the financial assessment.

This change of emphasis away from price has increased the standards on site and reduced the contractual conflict which often occurred.

12.8 Execution of the work

Having spent weeks or months developing and tendering the project and having secured finances and a Contractor, the Client is usually in a great hurry to get things moving.

One sure way of speeding up the start of the project is to issue the Contractor with a Letter of Intent. This is, in simple terms, confirmation of the Clients intention to enter into a contract with the Contractor pending completion of formal documents. The precise status of the Letter of Intent depends upon the wording and possible discussions and correspondence that may have taken place at the time of issue. The Contractor should satisfy himself that he is adequately safeguarded. A Contractor who goes over an upper financial limit or goes outside the wording of the Letter is likely to be exposed to financial risk.

The situation is by no means black and white. However, 99 times out of a 100, all goes smoothly, the formal contract takes over and the Letter of Intent falls away.

Another way of speeding up the process is to pressurise the Contractor to start quickly and with a reduced period for planning. Contractors should resist this as much as possible as this is one of the most critical stages for the project. It gives the Contractor time to select his best team and move resources around, select the best subcontractors and obtain the most commercially advantageous agreements. Furthermore, it allows the Contractor sufficient time to plan and deal with the vast number of Health, Safety and Environmental issues. Getting to site quickly may provide a short-term advantage but it rarely provides a long-term advantage and some aspects of the project may suffer. Clients can sometimes be quick to forget the reduced period given for planning and blame the Contractor when things go wrong as a result.

Once the Contractor is in receipt of an order or Letter of Intent the project team is selected. The Contractor may have been given an informal indication that he was to secure the project and some thought would already have been applied to this.

Once the project team is in place, work then starts on generating a number of documents. Some documents are required by law, some by the Contractor's Procedures and Management System and some by the Client as set out in the Contract Documents. Such documents include, but are not limited to, the following:

- Method statements
- Health and Safety risk assessments
- Environmental Aspects and Impacts assessments
- Health and Safety Plan
- Quality Plan
- Inspection and Test Plan
- Programme
- Cash flow forecasts
- Waste Management Plans
- CoSHH (Control of Substances Hazardous to Health) Assessments.

All these documents take a considerable time to prepare and set out the method of delivering the project in a safe and timely manner. These documents should be referred to regularly and updated accordingly as the project develops and should not just sit in the bottom of a drawer or filing cabinet.

Not only is it crucial that the Management Team has a full understanding of the project requirements but it is just as important that the entire supply chain and operatives are fully aware.

The relevant information for each activity is disseminated to the workforce by the site management team by means of:

- Site-specific inductions
- Method statement briefings
- Toolbox talks
- Workforce consultation.

Going through these once at the beginning is not enough. They should be repeated at regular intervals and each briefing should be recorded in writing so that periodic checks can be made to ensure compliance.

A competent Contractor would have a Health, Safety, Environmental and Quality Management System in place. Such a system would set out a number of control measures to ensure that the project is being delivered in accordance with the Contract and client requirements.

Some of these control measures include the following:

- Implementation of Quality Plan
- Implementation of Inspection and Test Plan

- Periodic Health, Safety, Environmental and Quality (SHEQ) inspections from the Contractor's Quality Manager, Environmental Manager and Health and Safety Manager
- Announced and unannounced visits from Senior Managers and Directors
- Announced and unannounced visits from External SHEQ Auditors
- Weekly Project Review Meetings including Safety and Commercial Reviews.

12.9 Summary

A successful contract is a safe contract where the whole project team has worked in unison to deliver the project on time and within budget. It requires the following.

- A realistic time-frame to deliver the project
- A detailed investigation of the concrete to identify the nature and extent of the problem
- A clear specification and schedule of works using BS EN 1504 and the CRA Method of Measurement for UK contracts
- Careful selection of materials and contractors
- Standard Conditions of Contract without amendment and a clear understanding of roles and responsibilities
- A strong emphasis on Health and Safety dealing with project-specific issues
- Quality control.

The tools are now available to achieve a successful outcome for contracts which involve concrete repair and provide a long-term solution to durability issues.

13 Sprayed concrete for repairing concrete structures

Graham Taylor

13.1 Introduction

Sprayed concrete was first used over a century ago, mainly for structures where formwork and normal placing methods were not possible. During the Second World War its potential for repair and rehabilitation was realised.

Since that time respect for it has grown as more research has been undertaken, new materials and equipment have been developed and contractors, banding together in trade associations, have worked towards reassuring their potential clients of the integrity of sprayed concrete.

Spraying concrete involves a fine-aggregate concrete, made with small-sized aggregate, projected at high velocity onto a hard surface. Impelled by compressed air or pump pressure, the material is rapidly placed and compacted and, with a high cement content and a low w/c ratio, its potential strength is high.

European harmonisation has required the system to be called sprayed concrete but the original process was patented as 'gunite' and this has been used as a generic term in the UK, whilst in America it is called 'shotcrete'.

13.2 History

The orginal Cement Gun was used by Dr Carlton Akeley, Curator of the Chicago Field Museum of Natural Science, in the late 1890s, to apply mortar over skeletal matrices to form replicas of prehistoric animals because the results of trowel-applied mortar were not acceptable.

The original cement gun, with a single chamber pressure vessel, into which a mixture of sand and cement was placed and then compressed air was applied, was the forerunner of today's twin-chamber dry mix method; the mixture travels down a hose under pressure, passes through a spray of water and then exits through a nozzle and then into place.

Gunite, as it was then known, came to Britain in the 1920s and healthy competition between rival firms gave rise to technical developments. During the Second World War, gunite, like everything else, was used as an expedient. One notable use was that basements of bombed buildings were cleared of rubble and then lined with gunite, in very short time spans, to produce water tanks for fire fighting.

After the war there was much to be repaired and gunite was used extensively as it gave a fast and effective answer to the problems of damaged and deteriorated jetties, warehouses, factories, grain silos, water towers, bridges, etc. It was also realised that gunite could be used as a strengthening medium, not merely as a means of repair.

Modern dry process machines do not have pressurised chambers but have a hopper, open to the air, into which the mixed material is fed. It then falls into one of a circle of chambers where it is pressurised with compressed air, and hence down the hose. These rotating barrel guns are almost constant feed-and-supply machines. In the 1960s the wet process was introduced, where cement, sand and water are mixed together before being put into the machine. Materials for repair are now often supplied pre-bagged and include various carefully selected sands and admixtures which give additional properties, such as non-shrinkage, and make the process more reliable.

Over the past forty years we have also seen the development of several codes of practice, specifications and methods of measurement, now largely superceded by the EFNARC specification and European standards.

13.3 Definition of sprayed concrete

Sprayed concrete is a mortar or concrete pneumatically projected at high velocity from a nozzle to produce a dense homogenous mass. It often incorporates a combination of admixtures and additions or fibres.

The material, which has zero or very low slump, is compacted by the force of the jet impacting on the surface, and can support itself without sagging whether on a vertical, sloping or, within certain limits, an overhead surface.

Rebound, material which bounces off the surface, is inevitable when concrete is projected at a relatively hard surface at high speed. It consists mainly of the larger aggregate particles, and should be discarded because it is of unknown grading.

13.4 The wet process

Definition

Cement, aggregate and water are batched and mixed together and then fed into a purpose-made machine. The mixture is then conveyed through a hose to a nozzle where the mixture is pneumatically and continuously projected into place. This can be seen diagrammatically in Figure 13.1. The mixture normally incorporates a set accelerating admixture and possibly other admixtures, and may also contain additions and/or fibres.

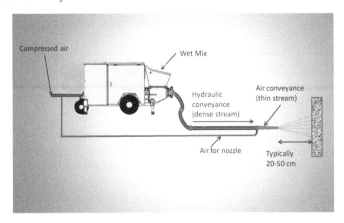

Figure 13.1 Diagram of a typical wet process system.

General

When blended bagged materials are being used, they are mixed with water on site and fed into the spraying machine. For large quantities, the mortar can be supplied ready-mixed or from a site mixer.

The mixing water can be accurately controlled and, with the use of water-reducing plasticisers, a w/c ratio of less than 0.45 is easily achievable. Sprayed concrete is a cement-rich material, although the surface area of the sand is high. Cement contents are normally in the range 350–450 kg/m^3 and 28-day strengths can be expected to be 30–60 MPa.

It is normal for the sprayed concrete contractor to design the mix so that he can take into account the performance of the wet concrete when it is being pumped and when it is in position, such as minimising the rebound, to produce a mix which will also meet the specified requirements of strength and durability.

With this process, rebound is less than that produced by the dry process.

Equipment

The wet process uses either piston or worm pumps to deliver the concrete to the nozzle in a dense stream. Compressed air is introduced at the nozzle to project and compact the concrete onto the substrate. A set accelerator may also be added at the nozzle to give the in-place material an early mechanical strength, which can minimise sloughing off.

For the low output normally required for repairs (up to 4 m^3/h) a worm pump would be suitable and this can take a maximum aggregate particle size of 4 mm. This output may seem very low to the conventional concretor but it produces an acceptable output of sprayed concrete. A double piston pump (in the range 4–25 m^3/h and capable of taking 20 mm aggregate), would be used for larger applications.

The entire spraying system should be able to deliver a constant stream of concrete free from pulsations which can give rise to segregation of the mix and over-dosing of the admixture(s).

The lines, of either flexible hoses or steel pipes, should be laid as straight as possible. They should be of uniform diameter and well sealed. The actual size used is dependent on the maximum aggregate size or the inclusion of a fibre in the mix. Prior to concreting, the lines should be lubricated with a grout.

All equipment should be cleaned and maintained to prevent the build-up of set concrete in the delivery system.

13.5 The dry process

Definition

Here only the aggregate and cement are batched, mixed together and fed into a purpose-made machine in which the mixture is pressurised, metered into a compressed air stream and conveyed through hoses or pipes to a nozzle, where water is introduced as a very fine spray to wet the mixture, which is then projected continuously into place, all as shown in Figure 13.2. Additions and/or fibres may also be in the mixture.

General

Water added at the nozzle needs only to be sufficient to hydrate the cement and give some fluidity to the mix. This produces a mix with a w/c ratio potentially as low as about 0.35 and consequently zero slump, enabling it to be placed without admixtures on vertical and overhead surfaces.

Should they be required, admixtures can be added, either dry in the mix or wet at the nozzle. Fibres can, of course, be added to the dry ingredients.

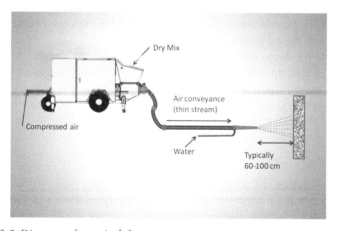

Figure 13.2 Diagram of a typical dry process system.

Spraying angles   Shotcreting angle

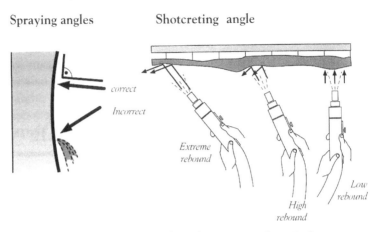

Figure 13.3 Effect of spray angle on rebound (courtesy Bekaert Ltd).

The nozzleman controls the water addition and whilst this at first seems somewhat unprofessional, if too little water is added the amount of rebound increases, and if too much water is added, the concrete will not have sufficient rigidity to stay in place. The angle of spraying is also critical to avoid excessive rebound (Figure 13.3).

When using a basic sand/cement mixture, rebound is higher than for the wet process; typically 10–15% when spraying downwards and as much as 50% when spraying overhead. Because this rebound is mainly the larger aggregate particles, the mix as-placed will be richer than the as-mixed materials, which would have an aggregate/cement ratio in the range 3.5 – 4.5 by weight. The resulting 28-day strength is typically 40–50 MPa. Pre-bagged materials, which contain cohesion-promoting admixtures, do not produce as much rebound. Figure 13.4 shows the spraying technique to achieve best quality, so it can be seen that a combination of spraying at approximately right angles and with a circular motion is preferred. However, to avoid shadow zones, it is also necessary to spray behind reinforcement, at an angle, when some rebound is unavoidable.

Concrete produced by the dry process can be the more variable. For example, because it is built up in layers and rebound is high, it can have variable density, especially at the interface between two layers. Rebound can also become trapped in corners, reducing the density. However, shrinkage is generally lower than for other similar but wetter concretes because of its low w/c ratio.

Equipment

The twin-chamber machines referred to earlier have now virtually ceased to be used and have been replaced by the rotor type, shown in Figure 13.5, where the dry materials are fed into an open hopper, which feeds a rotating barrel. As the barrel rotates, the pockets are brought under a compressed

Small circular motion

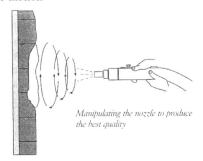

Manipulating the nozzle to produce the best quality

Figure 13.4 Spraying technique (courtesy Bekaert Ltd).

Figure 13.5 Typical dry spray machine.

air supply which projects the material down the hose. These machines, sometimes referred to as guns, have outputs up to 10 m³/h.

The length of hose is important. Short hoses will give rise to a pulsating effect in the material being delivered to the nozzle but this tends to even out in longer hoses. Whilst pressure, and the resulting exit speed from the nozzle, reduces with distance from the gun, hose lengths of up to 600 m are possible. With moderate hose length, the exit speed can be up to 125 mph.

13.6 Constituent materials

All materials must comply with the standards in force at the time. Sprayed concretes should contain cements, cementitious additions and aggregates or sands that have been proved to be suitable for this type of work. Sprayed concrete contractors have found by experience that cements of particular types and from particular sources work best for them.

Preliminary tests are normally carried out to establish the suitability of the constitutent materials and the final product. The term 'suitability' covers all aspects of the process as well as the final product's physical characteristics.

Cement

Most types of cement are used, as in any concrete work, depending on the requirement of the finished product. The cement to be used should be fresh and free from lumps.

Aggregates

Most sprayed concrete uses sand but aggregates up to 10 mm in size may be used where the section thickness allows, and provided the machinery will accommodate them.

The grading of the aggregate is important and should be within the limits shown in Table 13.1. The finer region is more suitable for dry mixes, although a high percentage of particles < 0.25 mm can lead to dust problems. Sand for finishing coats (sometimes referred to as flash coats as they are flashed over the surface as a sort of render) may be finer than the grading limits shown. Finer sands will generally produce a concrete with greater drying shrinkage while coarser sands will give more rebound. The aggregate size will also be restricted by the thickness of the repair section or coating.

In the dry process, the aggregate and sand need to be slightly damp to help suppress dust and to guard against segregation of the cement and sand in the hoses. The moisture content needs to be constant at around 5–6%. Too much moisture promotes clogging or blocking of the machine.

Table 13.1 Sand grading (Sprayed Concrete Association, 1999).

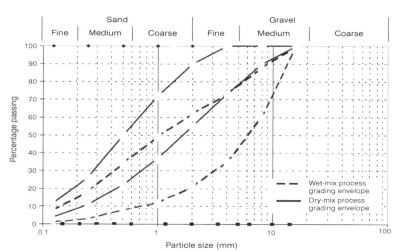

Mixing water

The requirements for mixing water are the same as for conventional concrete.

Admixtures

Admixtures can add significant flexibility to wet-process sprayed concrete and can overcome some potential difficulties, such as poor stop/start flexibility. A water-reducing admixture can be used to reduce the w/c ratio (and so increase the strength and durability) whilst maintaining the required workability and pumpability. Stabilisers (sometimes called 'hydration suspenders') can extend the workable life of the concrete, while accelerators are added at the nozzle to initiate setting once the concrete has been sprayed. However, using admixtures costs more and they can be caustic in nature and may even be detrimental to some properties of the sprayed concrete (e.g. accelerators can reduce long-term strength). All admixtures are complex chemicals and advice should always be sought from the manufacturer on dosage and application rates. Admixtures can alter their behaviour when added in combination, so care should be taken to ensure compatibility. Admixtures for sprayed concrete should comply with the requirements of BS EN 934-2 and BS EN 934-6 or EFNARC (1996). BS EN 934-5:2007 covers admixtures for sprayed concrete.

Superplasticisers and water reducers

Superplasticisers are high-performance water reducers which disperse the fines more effectively within a mix and therefore improve workability and cohesion. They need to be used with care and trials are recommended to determine the optimum dose and properties. The commonest types are sulfonated melamine formaldehyde condensates (which form a lubricating film on the particle surfaces) and sulfonated napthalene formaldehyde condensates (which electrically charge the cement particles so that they repel each other), although many hybrids now exist. Water reducers are commonly based on lignosulfonates or hydroxylated carboxylic acids and are utilised in a similar way to superplasticisers. The dosage of superplasticiser and water reducer depends on the mix specification, w/c ratio, required workability, and cement and aggregate types.

Superplasticisers and water reducers can reduce the water/cement ratio for a given workability (thereby improving the compressive strength and other properties) or increase the workability for a given w/c ratio (thereby increasing the pumpability of a mix). The workability will usually return to normal approximately 20 minutes after the addition of the superplasticiser (depending on the type) and so it should be added immediately before placing. They do not adversely affect the final hardened properties of the concrete, although the setting time may be increased slightly.

Air entrainment

In freezing conditions, the pore moisture in concrete can freeze and the resulting expansion of the ice can cause scaling and delamination of the concrete surface. This is called freeze–thaw damage and is distinct from frost damage, which occurs when the water in concrete which has not yet hardened, freezes. Air-entraining agents are added to wet-sprayed concretes to create a hardened concrete with small, well-distributed air pores, consequently improving freeze–thaw durability and de-icing chemical scaling resistance. The ACI Guide to shotcrete 506R (2005) specifies that wet-mix shotcrete exposed to severe freeze–thaw conditions must be air-entrained, but up to half the air content can be lost on impact during the spraying process (Nordstrom, 1996). It can therefore be difficult to achieve an *in situ* air content above 4% although there is some evidence to suggest that sprayed concretes require a lower air content (2–3%) for freeze–thaw resistance than conventionally placed concrete (Austin and Robins, 1995). Airentrainment can also reduce the flow resistance in a pump line, to some extent acting as a lubricant between the particles. The entrained air is then forced out as the mix impacts on the substrate, thus stiffening the mix. However, air entrainment can sometimes cause difficulties when pumping under high pressure as the bubbles can be compressed or destroyed, increasing the possibility of blockage. A pre-construction pumping trial should highlight any problems with the air entrainment.

Polymers

Latex solutions (such as styrene-butadiene rubber, SBR) are added to improve bleeding resistance, pumpability and adhesion. It is also claimed that they can improve the tensile and flexural strength, permeability, abrasion and chemical resistance of the hardened concrete (Warner, 1995). Research has recently been completed on the addition of polysaccharide gums in sprayed concrete, which appear to produce a balance between pumpability and sprayability by reducing apparent viscosity at high shear rates (e.g. in the mixer), but have less affect at low shear (i.e. after spraying) and hence allow good build (Ghio and Monteiro, 1998).

Accelerators

Accelerators are rarely required in low-volume applications but can be essential when rapid strength development is necessary. They are available in powdered form (mainly for dry spraying) or liquid form (generally added at the nozzle). It is important to keep the accelerator dosage as low as possible to achieve the desired results as strength and durability reduce with age and dosage level. Accelerators can reduce the 28-day strength of control mixes without accelerators. Prediction of the setting time and strength gain is difficult as they are dependent on the types and sources of both the cement and the accelerator. Therefore, trials should be conducted each time an accelerator is to be added, either at a different dosage level or with a new mix design. There are a variety of types, the more recent being low-caustic to minimise adverse health effects.

Retarders and stabilisers

Traditional retarders are not common in sprayed concrete work. However, stabilisers are often incorporated (in conjunction with an activator added at the nozzle) to control hydration where long periods of standing-time are expected in the spraying process, such as in underground work where the concrete is batched above ground and then conveyed large distances to the spraying pump.

Fibres

Both steel and synthetic fibres are used to satisfy a performance requirement of the finished concrete. Steel fibres, which may be included for abrasion resistance, have the ability to cause blockages in pipes and hoses, hence the general requirement to keep their length down to not more than 0.7 of the pipe/hose diameter until it has been proved that longer fibres can be sprayed without causing blockages. Synthetic fibres, used, for example, to control cracking, are flexible and, provided they are well-distributed, flow easily through the hose.

Cementitious additions

Silica fume, pulverised fuel ash (pfa) and ground granulated blastfurnace slag (ggbs) can all be used in sprayed concrete; generally up to the percentages given in Table 13.2.

Condensed silica fume, microsilica or silica fume, as it can be known, is now commonplace in sprayed concrete. It is formed from the production of silicon or ferrosilicon alloys and the average particle size is typically 0.15 microns. It has a high SiO_2 content and is a very reactive pozzolan. This extremely fine material is available in powder form for the dry process, or in wet, slurry form for wet process mixes.

Table 13.2 Cementitious additions

Material	Max addition (% of Portland cement)
Silica fume	15
pfa	30
ggbs	30

Inclusion into the mix at between 5 and 10% provides a wide range of performance enhancements. Firstly it increases the cohesiveness of the fresh concrete and improves the bond of the sprayed concrete to the proposed surface. This cohesiveness also has a big impact on rebound, reducing it from about 40% to less than 10% on typical mixes. Any 'rebounded' material is unusable once sprayed so considerable cost savings can be made. It also reduces the amount of dust created by the spraying action. The cohesiveness of the mix then improves the build-up characteristics, allowing increased thicknesses of sprayed material. Collectively this reduced dusting and lower rebound enables more efficient production and improved working conditions.

For wet processes, improved pumpability is noticeable. Silica fume is often added to conventional concrete as a pumping aid and this enhancement is also true for sprayed concrete.

A number of benefits accrue when using silica fume. These include:

- Increased compressive, flexural and tensile strength and increased bond to the surface and reinforcement,
- Reduced permeability,
- Improved frost- and chemical-resistance.

Tests have shown strengths in excess of 100 MPa can be obtained using a quality silica fume. Similarly, early-age strengths of 1 MPa at only 2 hours can be achieved using accelerators. The natural increased cohesiveness of a silica fume mix can also have an impact in lowering the required accelerator additions.

For special applications higher dosages can be considered and suitable trial mixes undertaken to test suitability. Major suppliers will be able to advise on mix details, provide sample material for testing and a range of site support services (storage, pumps, etc.) if required.

The quality standard for silica fume is BS EN 13263 (BSI, 2005) and this should be considered as mandatory when considering a likely source.

13.7 Spraying procedures

As a bond is required in the case of repairs or strengthening work, the substrate must be clean before spraying of concrete can commence. If the substrate surface is porous it may need wetting down to kill excessive suction from the concrete being sprayed. In dry spraying the w/c ratio of the sprayed concrete is already very low and suction from the interface surface might reduce it to below the critical level for complete hydration of the cement. However, there should be no free water left on the surface before spraying commences.

Before applying dry sprayed concrete, the nozzleman should direct the spray away from the works until he judges that the concrete coming out of the nozzle is of an acceptable quality.

Layers are built up by making several passes of the nozzle over the area, generally in an overlapping oval configuration. The nozzle should be at right angles to the substrate and at the correct distance from it. This would be between 600 mm and 1 m for the dry process and between 250 and 500 mm for the wet process. The nozzleman will recognise the correct distance by the amount of rebound and the effectiveness of compaction (Figure 13.6).

Thickness of the section to be sprayed can be indicated by stretched 'piano' wires located at the required finished surface. Application should start at the bottom of a vertical or sloping surface and from the shoulder to the crown for overhead working. As repairs usually contain corners and sharp angles, care should be taken to avoid trapping rebound material at these critical locations.

The nozzleman's assistant should remove any rebound that might become encased in the works and this rebound should be discarded, not fed back into the system, as its grading has changed. He may also be required to trim back the sprayed material to the line indicated by the stretched wires, although it is better practice to spray to a thickness less than that indicated and to finish off with a flash coat, which is more like a render than a concrete. Trowelling or screeding is best avoided as this action creates tension cracks in the sprayed concrete's surface.

Once a section is complete, it must receive curing, which can be a sprayed-on membrane (if no further coating is to be applied) or any of the other conventional curing techniques. Unfortunately, where large areas are

Figure 13.6 Concrete spraying.

involved, proper curing is often omitted and the result can be premature drying out, excessive shrinkage or delamination.

13.8 Quality assurance

Introduction

Sprayed concrete is more dependent on those who place it than is conventional concrete; a trained nozzleman is therefore imperative (Figure 13.7). This also makes it difficult to separate quality control from quality assurance. The nozzleman controls the water in the dry process and, as a result the *in situ* mix proportions will be different from the batched proportions. If the requirement is that it should adhere to the substrate, it is the nozzleman and his assistant whose actions will control whether or not it does. The nozzleman's method of working may create variations in the quality of succeeding layers and therefore density (or perhaps even homogeneity), as well as strength, needs to be checked. Cores are used for this purpose and these can also help to check the thickness of the cover to the reinforcement.

Prequalification of nozzlemen

Each nozzleman should demonstrate his skill prior to commencement of spraying and this is usually done by spraying into test panels, usually about 750 × 750 × 100 mm thick, set at the orientation of the proposed works (vertical, sloping, horizontal or overhead). The test panels should replicate the type of concrete, i.e. reinforced (bar, fabric or fibres) or unreinforced.

Cores are subsequently taken to ensure that strength, density, freeze/thaw resistance, chemical resistance and filling behind reinforcement are all acceptable.

13.9 Quality control

Quality control must be continuous throughout the works, from drawings through to finished product and must be carried out by someone with experience of this type of work. The following should be included:

Review

The method of placing the concrete must be considered whilst the type, size and location of the repairs are being determined. Reinforcement is a main consideration – bar sizes, locations and staggering of overlaps, methods of (rigid) support are all important. Large bars, laps and small gaps behind exposed bars can all cause shadowing; that is, where the sprayed concrete behind the bar does not get fully compacted.

Figure 13.7 Nozzleman at work.

Potential debonding and delamination

To detect the presence of debonding and delaminations, the concrete surface is tapped with a steel hammer some 2 to 3 weeks after placing. Sound concrete will ring, whilst debonded materials will give a dull sound. A hand placed on the surface during the tapping will help to detect the extent of any debonding.

Tensile bond strength

Whilst this is not a perfect test and can only tell the strength at one particular location, it is widely used to test the interface bond strength or the tensile strength at the weakest point. The test may need interpretation.

To perform this test, a core bit is used to drill just into the underlying layer, which might be a substrate or a previously sprayed concrete layer. A steel disc is attached to the top of the core using epoxy resin and a pull-off load is applied to the disc through a loading rig and a hydraulic ram, until failure occurs. Any deviation of the direction of the force from the perpendicular can induce bending which will give very misleading results. The failure plane is then examined to determine if it has occurred at the interface or within the body of the sprayed concrete, or even within the substrate. A typical pull-off tester is shown in Figure 13.8.

Figure 13.8 Typical pull-off tester (courtesy of Safe Environments, Australia).

Strength and density

Cores taken from the placed material can be tested as for normal concrete. Although the core diameter is usually small, the aggregate is also smaller than for conventional concrete and specimen size and proportion allowances can be made.

In situ testing

There are various tests which can be carried out *in situ*.

Entrapments of sand or rebound material, if suspected, can be verified by taking cores. Ultrasonic pulse velocity testing (Pundit) can be used to indicate strength and thermography or x-rays can be used, if rarely, to evaluate the condition of the sprayed concrete.

Daily test panels

Whilst these are normally specified, to test mainly for compressive strength and density, they may not represent the concrete in the works. The nozzleman obviously knows what test panels are and what they are used for, so he ensures that only the best concrete and workmanship go into them. The quality of concrete at start-up is frequently below the required standard.

If cores from these daily test panels are found to be sub-standard, then cores should be taken from the *in situ* concrete sprayed on that day and tested as required.

13.10 Conclusion

Whilst the above may indicate difficulties in controlling the quality of the end product, the material has a higher than normal cement content and the spraying process will only allow good material to pass through the system. Problems, if any, are often associated with the skill of the nozzleman and it is therefore imperative that only well-trained, competent ones are used.

13.11 Applications of sprayed concrete

The Sprayed Concrete Association's book *An Introduction to Sprayed Concrete* starts with the words 'Sprayed concrete is a versatile and economic method of concrete placement, with endless possibilites ...' and, yes, the list of actual and potential applications is almost limitless but it succeeds most notably where the use of rigid formwork is absent, e.g in strengthening and repairing:

- Sea and river walls
- Aqueducts
- Water towers
- Canal linings
- Ditches
- Concrete damaged by reinforcement corrosion
- Fire-damaged structures
- Cooling towers
- Bridges
- Jetties and wharves
- Brick arches and tunnels.

The Concrete Society's Technical Report No 56 (Concrete Society, 2002), details a wealth of material on the use of sprayed concrete in repair. A typical set of repair scenarios is given in Table 13.3.

Acknowlegements

The author and editor would like to thank the Sprayed Concrete Association, in particular Ken Dykes, for helpful comments in the preparation of this chapter.

Table 13.3 Typical uses for sprayed concrete

| Repair scenario | Purpose | Geometry | | | Substrate | |
		Size	Depth	Tolerance	Type	Surface characteristics
Bridge soffits	Cover for durability Structural	< 2 m³	50–100 mm	±10 mm	Concrete	Carbonation Chlorides AAR
Bridge abutments, marine structures	Cover for durability Structural	< 2 m³	50–100 mm	±10 mm	Concrete	Carbonation Chlorides AAR
Buildings, water retaining structures, RC chimneys, car parks	Cover for durability Structural	< 2 m³	50–100 mm	±10 mm	Concrete	Carbonation Chlorides in car parks
Fire-damaged structures	Structural	< 2 m³	50–100 mm	±3 mm (visible) ±10 mm (covered)	Concrete	Fire damage
Tunnels	Structural	< 1 m³	100 mm	±10 mm	Concrete	Carbonation Chlorides
Sewers, masonry tunnels, arch bridges etc.	Strengthening	1+ m³	25–50 mm (< 100 mm)	±10 mm	Masonry	Deteriorated masonry

Preparation	Environment	Orientation	Reinforcement	Surface finish	Access
Hydro-demolition + grit blasting	Atmospheric	Overhead	Mesh fibre	Trowelled finish no colour match	Limited to night working Restricted space
Hydro-demolition + grit blasting	Atmospheric	Vertical	Mesh fibre	Trowelled finish no colour match	Limited to night working Restricted space
Hydro-demolition + grit blasting	Atmospheric	60:40 vertical: overhead	Mesh	Trowelled finish colour match (where no surface coating provided)	External repairs use scaffold, platforms etc.
Hydro-demolition + grit blasting	Atmospheric (substrate absorbs water at high rate)	50:50 vertical: overhead	Mesh (replace damaged steel)	Trowelled where visible, otherwise as shot, no colour match	May be hazardous
Hydro-demolition + grit blasting	Cool (ventilation fans) can be damp	Overhead	Mesh (replace corroded steel)	As shot, no colour match	Restricted access to road and rail tunnels Pumping long distances
	Warm and damp	Curved surfaces	Mesh (stainless steel)	As shot, no colour match	Restricted access through manholes Pumping long distances

References

American Concrete Institute, *ACI Guide to Shotcreting*, Committee Report 506R, 2005.

Austin, S.A and Robins, P.J.R, *Sprayed Concrete: Properties, Design and Application*, Whittles Publishing Services, Latheronwheel, UK, 1995.

British Standards Institute, BS EN 934-2:2009, 'Admixtures for concrete, mortar and grout. Concrete admixtures. Definitions, requirements, conformity, marking and labelling', BSI London, 2009.

British Standards Institute, BS EN 934-5:2007 'Admixtures for concrete, mortar and grout; Part 4: Admixtures for sprayed concrete – Definitions, requirements, conformity, marking and labelling', BSI London, 2007.

British Standards Institute, BS EN 934-6:2006: 'Admixtures for concrete, mortar and grout; Sampling, conformity control, evaluation of conformity, marking and labelling', BSI London, 2006.

British Standards Institute, BS EN 13263:2005 'Silica Fume for Concrete', BSI London, 2005.

Concrete Society, *Construction and Repair with Wet-process Sprayed Concrete and Mortar*, Technical Report No. 56, The Concrete Society, 2002.

EFNARC, *European Specification for Sprayed Concrete*, European Federation for Specialist Construction Chemicals and Concrete Systems, 1996.

Ghio, V.A. and Monteiro, P.J.M., 'The effects of polysaccharide gum additives on the shotcrete process.' *ACZ Materials Journal*. Vol. 95, No. 2, March/April, 1998, pp.152–157.

Nordstrom, E., 'Durability of sprayed concrete – a literature study', in R.K. Dhir, and P. C. Hewlett (eds) *Radical Concrete Technology: Proceedings of the International Congress, Concrete in the Service of Mankind*. Dundee, June 1996. E&FN Spon, London, 1996. pp.607–617.

Sprayed Concrete Association, *An introduction to Sprayed Concrete*, 1999.

Warner, J., 'Understanding shotcrete – the fundamentals', *Concrete International*, Vol. 17, No. 5, pp. 59–64, 1995.

14 Durability of concrete repairs

G. P. Tilly

14.1 Introduction

The inspection, maintenance and repair of concrete structures absorb a considerable volume of the construction budget in the UK and elsewhere. In Europe it has been estimated that it is some 50 per cent of the total and in the United States the annual expenditure due to corrosion of reinforcement alone amounts to some 8.3 billion dollars. Moreover, in extreme cases of deterioration there have been structural collapses (Doyle, 1993, Oliver, 2000, Government of Quebec, 2007). In consequence of these problems it has been recognised that concrete structures must be more durable in the first place and repairs must be effective.

Repairs to damaged or defective concrete have been the subject of much research at academic institutes and in the repair industry. The professional institutions and code-writing bodies have published state-of-the-art advisory documents and guidelines to inform the workforce and raise the quality of repair work. The construction industry has developed improved materials and techniques to enable repairs to be more effective and durable in the long term. Unfortunately these efforts have not been entirely successful as it is generally found that the performance of concrete repairs is rather disappointing. It has been difficult to express performance quantitatively as there have been few surveys of repairs and those that have been reported have been either localised or lacking in sufficient detail or numbers to enable useful conclusions to be drawn. Surveys of structural condition are fairly common and have usually been carried out in response to concern about a particular condition (Radic, 1989, Rowe et al., 1984, Van Begin, 1987). Reports on the durability of repairs are usually based on laboratory work whereas fieldwork is much less common. In an overview of the durability of repair materials Cusson and Mailvaganam (1996) noted that:

> The lack of comprehensive data on the performance of repair products and on the potential incompatibility between repair materials and substrate concrete is at least partly responsible for the large number of premature repair failures in North American concrete structures.

A survey of performance of repairs, by the US Corps of Engineers (Emmons, 2006) reported that only 50 per cent of repairs to their structures were classified as good, 25 per cent were considered fair or poor, and 25 per cent had failed. Unfortunately no other information was provided.

The owners and managers of concrete structures have varying requirements but all need to have reliable knowledge of the durability and performance of any repairs that are undertaken. For industrial installations having high risk to the general public, such as nuclear plant, it is essential for the structures to be of wholly durable concrete having long life and no conceivable possibility of failure. At the other end of the spectrum there are commercial structures where requirements are short term and 'quick fixes' are acceptable provided they are safe. National authorities such as highway agencies have high expectations of repairs and invariably require construction and maintenance to be carried out to their own standards. Some require the repairs to be supervised and inspected by their own in-house staff.

The appearance of repairs can be contentious. In some cases structural integrity is all-important and unsightly repairs are acceptable within reason and provided they meet requirements of integrity and durability. In others, the repairs are required to be seamless and comparatively minor differences in material, texture or colour can be unacceptable. This applies to some of the older structures that have become classed as having historic importance (Tilly, 2002a).

Against this background an EU-funded project, CONREPNET, was carried out with the objective of developing the concept of performance-based rehabilitation of reinforced concrete structures. The outcome of this project is described in two publications (Matthews et al., 2007, Tilly and Jacobs, 2007). The first step in this project was to collect and evaluate case histories containing data relating to the performance of concrete repairs. This involved the evaluation of repair performance and identification of the modes and causes of premature failures.

This chapter addresses reasons for the disappointing performance of concrete repairs and ways to improve the situation.

14.2 Case histories

Case histories of concrete repairs were mainly supplied by members of the CONREPNET project, and UK data for cathodic protection installations were supplied by consultants. Others were obtained from the literature and first-hand experiences of the author. The suppliers were from all sides of industry and included consultants (23 cases), repairers (64 cases), academic institutions (83 cases) and owners of structures (60 cases).

The data were from Finland, Denmark, Sweden, Czech Republic, Germany, France, Belgium, the Netherlands, Spain, Greece and the UK, shown in Figure 14.1 by geographical area, and are representative of a wide range of European climates and conditions.

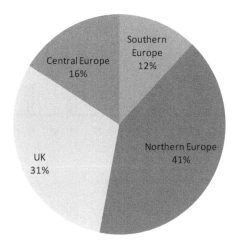

Figure 14.1 Case histories by geographic area.

A total of 230 full case histories were collected. Data were extracted and entered into a database to enable the trends to be analysed and reported. The data were concerned with:

* Information about the structure
* Initial causes of damage
* Type of repair
* Performance of repair.

Some of the case histories were incomplete, for example where construction dates were not known by the respondent. Case histories of 'successful' repairs carried out less than 5 years before inspection were included as examples of repair but excluded from analyses of performance as they were too recent to have experienced the full rigours of weathering and operational activities.

In addition to the full case histories, 62 abbreviated cases were obtained for repairs in the UK involving cathodic protection (CP).

The structures that were reported comprised 77 buildings (domestic and commercial), 75 bridges (road and rail) and 78 industrial structures (car parks, power stations, dams and other less common structures).

Ages of the structures at the time they were reported were between 151 years (the oldest) and less than 10 years. Altogether 51 structures were more than 50 years old but most (63 per cent) were between 20 and 50 years old, as shown in Figure 14.2.

The distribution of environments is shown in Figure 14.3. The most common is urban (37 per cent of the total) followed by highways (24 per cent) and rural (23 per cent).

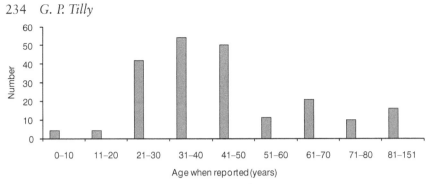

Figure 14.2 Ages of the structures when reported.

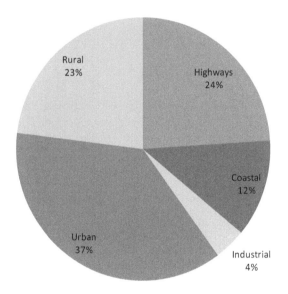

Figure 14.3 Distribution of environments.

14.3 Problems requiring repair

The case histories involved a variety of initial problems requiring the repair work. In some there was more than one problem reported, for example corrosion and consequential loss of strength through reduced section in the reinforcement bars. Corrosion, frost damage and cracking could also occur together. In these cases it was usually possible to identify the primary problem but in the case of corrosion and cracking it was difficult to decide whether cracking had led to the corrosion or vice versa.

Repair is defined as the restoration of the structure to its initial durability and strength. It follows that strengthening to a level above the initial design value is excluded from the case histories.

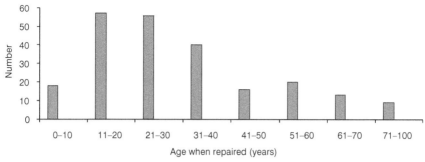

Figure 14.4 Ages of structures when repaired.

Ages of the structures at the time of repair are shown in Figure 14.4. Most (66 per cent) were in the range 10 to 40 years, 42 structures were over 50 years and the oldest when repaired was 100 years.

It is convenient to consider the initial problems under two groups: Deterioration involving underlying processes of degradation, e.g. corrosion and AAR, and Damage caused by mechanical or other actions.

14.3.1 Corrosion and AAR

Corrosion is by far the most common problem in concrete structures and occurred in 137 of the case histories. The corrosion was almost exclusively in the reinforcing steel but there were nine cases in prestressing steel. Corrosion of prestressing steel is potentially more serious than reinforcement as it can, and has on several occasions, led to collapse (Tilly, 2002b).

Corrosion invariably occurs as a result of a fundamental cause such as inadequate thickness of cover or leaking expansion joints. It follows that there are two aspects of the repair to be addressed: to take action to eliminate the root cause (e.g. by waterproofing), and to halt the internal process of corrosion (e.g. by removal of contaminated concrete).

AAR is less common and there were only 14 cases reported, five accompanied by corrosion. There is considerably less experience of AAR and the relative numbers of cases of corrosion and AAR are not unrepresentative of the overall numbers in the field. AAR can be very damaging and, like corrosion, has led to structures having to be demolished.

The processes of corrosion and AAR are difficult to stop and invariably continue unless the repairs are very carefully (and expensively) carried out.

14.3.2 Mechanical and other damage

This group of problems includes freeze–thaw damage (due to cycles of freezing and thawing), cracking caused by structural actions, wear and leaching (of spillways on dams), impact damage (by vehicles or ships striking the structure), vibrations and repeated loading causing fatigue damage, overloading, leakage of water-retaining structures and deflection caused by progressive creep.

Faulty construction is also included in this group except when it has led to corrosion or AAR. It is usually detected immediately after completion of the structure but seven of the cases were not identified until much later. In most of the reports faulty construction involved cracking but it was not clear whether it had been present since the structure was built or had developed later as a consequence of some other fault.

Altogether there were 78 case histories of mechanical and other damage. The distribution of the types of damage is shown in Figure 14.5.

14.4 The repairs

In general the repairs were designed and carried out with the objective of achieving durability for the remaining life of the structure but the more realistic expectation was somewhat less. There was only one case where it was stated that the repair was a 'holding action' to last a short time until the structure could be replaced. In practice temporary repairs may be carried out for a variety of reasons and it is not unusual for them have to remain much longer than intended.

The most common types of repair were: patching, coating, crack injection, restoration of strength (replacement of excessively corroded reinforcement bars, added prestress), sprayed concrete, and electro-chemical methods (CP, realkalisation, chloride extraction). Other less common methods included: application of corrosion inhibitors, bonded plating, wrapping with carbon fibre reinforced plastic (CFRP), and added cover concrete. In 60 per cent of the case histories more than one method was used, for example patching plus coating.

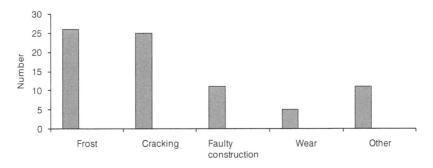

Figure 14.5. Distribution of damage requiring repair excluding corrosion and AAR.

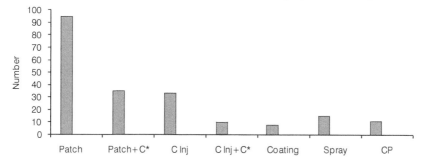

Figure 14.6 Types of repair to corrosion damage.

*+C denotes coating.

14.4.1 Repairs to corrosion and AAR

Corrosion

The most common techniques, and combinations, used to repair corrosion are shown in Figure 14.6 and summarised as follows:

- Patching was carried out in 70 per cent of the repairs to corrosion and 35 per cent of the patches were accompanied by coating.
- Crack injection was carried out in 25 per cent of the repairs and 30 per cent of the injections were coated.
- Sprayed concrete was applied in 10 per cent of the repairs.
- Solo coatings were applied in 5 per cent.
- CP was applied in 8 per cent.

In the nine cases where prestressing steel had corroded, eight required strengthening. Four were strengthened by added prestress.

AAR

There were 14 cases of AAR reported and five were accompanied by corroded reinforcement. Nine of the repairs involved patching, six of them being coated. The techniques are summarised in Figure 14.7.

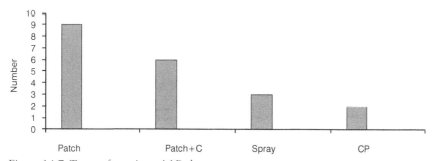

Figure 14.7 Types of repair to AAR damage.

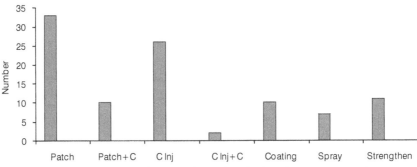

Figure 14.8 Types of repair where mechanical and other damage occurred.

14.4.2 Repairs to mechanical and other damage

The most common techniques and combinations used to repair damage caused by mechanical and other actions are shown in Figure 14.8 and summarised as follows:

- Patching was carried out in 40 per cent of the repairs and 30 percent of the patches were accompanied by coating.
- Crack injection was carried out in 30 per cent of the repairs and 10 per cent of the injections were accompanied by coating.
- Sprayed concrete was applied in 10 per cent of the repairs.
- Solo coatings were applied in 25 per cent.
- Strengthening was carried out in 15 per cent.

Note that due to some of the repair techniques being used in combination, the above percentages exceed 100 percent in total.

14.5 Performance of repairs

Performance of the repairs was classified by the respondents as follows:

- Successful and not requiring attention, as identified at the last inspection
- Exhibiting early evidence of failure and considered to be unsatisfactory but not requiring immediate attention
- Failed and requiring immediate attention.

Table 14.1 Performance of repairs to corrosion and AAR

	Successful
Corrosion	50%
AAR	20%

Table 14.2 Performance of repairs to mechanical and other damage

	Successful
Freeze–thaw	25%
Cracking	65%
Wear and leaching	80%
Faulty construction	80%
Other damage	45%

Classification was by no means unambiguous as to some extent it required the respondent to make a subjective judgement based on information provided by the inspection report. For example, a repair having minor cracking may be seen as being structurally successful by the repairer but considered to be aesthetically unacceptable, or exhibiting early signs of failure by the owner of the structure. These situations can lead to disputes but ultimately repairs have to be carried out to the of the owner and it is the responsibility of the repairer to clarify performance requirements beforehand.

14.5.1 Performance of all types of repair

For all types of repair, 50 per cent were reported to be successful, 25 per cent exhibited early evidence of failure and 25 per cent had failed. Interestingly, these data are identical to those of the US Corps of Engineers (Emmons, 2006).

Common requirements are for contractors to guarantee repairs for 5 or 10 years, depending on the situation. National authorities having their own

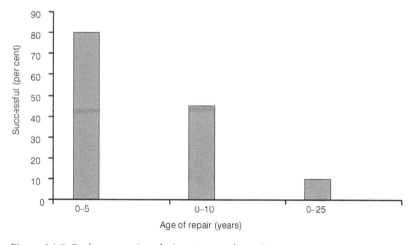

Figure 14.9 Performance in relation to age of repair.

standards have a general expectation that repairs should last for at least 25 years. Performances of the 230 case histories in relation to these requirements are shown in Figure 14.9 and it is evident that there is a shortfall in the expected performance.

The most successful repair lasted for 52 years and 10 per cent of the 230 repairs lasted in excess of 25 years.

14.5.2 Performance in relation to type of repair

Performances of the most common repairs, whether applied solo or in combination, to all types of damage, are shown in Figure 14.10.

Patches

The two materials mainly used for patching were cementitious and polymer modified mortars; others included fibre-reinforced mortars. Performances were as follows:

- Cementitious patches, 45 per cent successful
- Polymer modified, 50 per cent successful
- Other materials, 25 per cent successful.

Although polymer modified materials are generally considered to be more effective than cementitious mortars, there was little difference in the performances in practice. This may be a reflection of assessments based on laboratory work compared with behaviour when subject to the rigours of natural weathering in the field or it may be due to earlier materials being less effective.

Coatings

Several types of coating were reported: barrier, hydrophobic and others (aesthetic, anti-carbonation, unspecified). Performances were as follows:

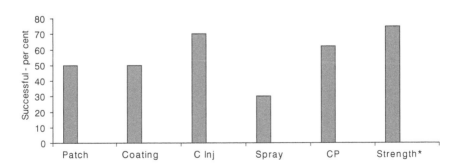

Figure 14.10 Performance in relation to type of repair.

* Restoration of strength excluding replacement of corroded reinforcement

- Barrier coatings, 50 per cent successful
- Hydrophobic coatings, 55 per cent successful
- Other coatings, 26 per cent successful.

The poor success rate for other coatings can be attributed to factors such as the use of aesthetic coatings in unsuitable situations.

The application of a coating over patches or injected cracks produced a marked improvement in durability:

- Patches were 30 per cent successful when solo and 50 per cent successful when coated.
- Crack injections were 70 per cent successful solo and 80 per cent successful when coated.

Crack injection

The high success rate for crack injection (70 per cent) may be because it was used in cases where corrosion was less advanced so that sealing the cracks was sufficient to stop the process.

Cathodic protection

The CP data comprised 12 fully comprehensive case histories and 62 abbreviated cases. The performances of the 74 installations were as follows:

- Wholly successful, 60 per cent
- Requiring attention, 20 per cent
- Failure, 20 per cent.

Restoration of strength

Restoration of strength was not always the main object of the repair as some 70 per cent of the problems also involved corrosion, many coupled with frost damage. In this analysis the cases involving corrosion alone have been excluded and the losses in strength were due to mechanical causes.

- 75 per cent of the repairs were successful.

14.5.3 Performance of repairs in relation to type of damage

Corrosion and AAR

It is evident from Table 14.1 (p. 238) that it is particularly difficult to halt the process of AAR and repair the damage successfully as only 20 per cent of the repairs were successful.

Mechanical and other damage

The data shown in Table 14.2 (p. 239) are for types of damage excluding cases where corrosion also occurred.

It is evident from Table 14.2 that repairs to mechanical damage were relatively successful. The poor performance of freeze–thaw repairs may be due to continuing damage to poor quality concrete which was used in the first place.

14.5.4 Modes of repair failure

The modes of repair failure are shown in Figure 14.11. The most common are continued corrosion, cracking and debonding. Other less common modes included deteriorated concrete, spalling, efflorescence, discoloured coating, wear and breakdown of CP. The modes of CP breakdown included:

- Failures of management (accidentally switching off)
- Electric failures (short circuit, overheating, failed control)
- Deterioration with age and weathering.

Failures of CP systems are generally less costly than others because they can often be rectified by simple measures such as switching on again, correcting an electrical circuit or replacing a control box. Also, the benefits of CP continue for some time after current ceases to flow so that there is enough time to repair the system before corrosion is re-established.

Failure modes of patches (the most common method of repair), were:

- Cracking, 40 per cent
- Debonding, 30 per cent
- Continued deterioration (corrosion, AAR, leakage), 30 per cent.

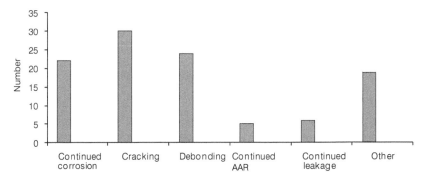

Figure 14.11 Modes of repair failure.

Figure 14.12 Freeze–thaw damaged underpass repaired with a sprayed polymer coating.

Figure 14.13 AAR repaired with polymer modified mortar and a polymer coating.

Failure modes of coatings applied solo included cracking, continued leakage, continued corrosion and deteriorated concrete. Where failed coatings were over patches it is not clear whether failure originated in the patch or the coating.

Case histories of failed repairs, for an underpass, a bridge and a lift tower, are illustrated in Figures 14.12 to 14.14.

The repair failure shown in Figure 14.12 was caused by the pressure of salt crystallisation causing cracks and debonding of the coating. It was concluded that the repair had been incorrectly designed and the wrong repair material was used. Moreover, the back of the wing wall had not been waterproofed.

The bowstring bridge shown in Figure 14.13 had deteriorated due to AAR. The repair failure was caused by access of water to the concrete which encouraged continued AAR. It was concluded that the problem had been incorrectly diagnosed and the wrong repair system was used.

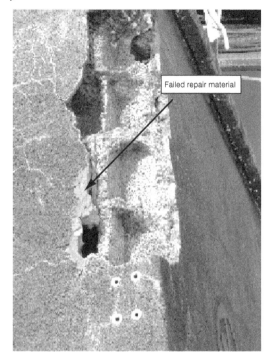

Figure 14.14 Damaged concrete repaired with incorrect material.

14.6 Causes of repair failure

In some cases the respondents admitted that the cause of failure was not known and in others the comments were rather vague. This is not altogether surprising as non-destructive testing was applied in only 15 per cent of the cases so that diagnoses were based on visual inspections alone. The main causes of failure were given as follows:

- Incorrect design of the repair, for example repairs to AAR which failed to address the root cause.
- Use of incorrect material, for example a patch composed of material that was much stronger than the substrate, selection of aesthetic coatings for application in aggressive conditions.
- Poor workmanship, including failure to follow manufacturer's instructions.
- Incorrect diagnosis of the underlying problem, for example dealing with corroded reinforcement but failing to stop leakage that caused the problem in the first place.
- Other causes included vandalism, the repairer failing to follow instructions, excess economy and extremes of weather during the repair work.

The incidences of the different causes of failure are summarised in Figure 14.15.

The geographic location and environment had a significant influence on performance of repairs. As expected, repairs in coastal and industrial environments performed very poorly. On the other hand it was surprising that repairs in an urban environment were more successful than those in a rural environment.

The influence of environment on repair performance is summarised in Figure 14.16.

14.7 Concluding remarks

The generally poor performance of concrete repairs comes as no surprise to practitioners. Moreover this is despite the volume of research, development of improved repair materials, and guides to good practice that have been produced.

Some of the trends highlighted by the survey and comments by the respondents are, however, helpful in indicating ways to make improvements in the future. There is a general view that pressures on costs cause repairers to take cheaper options, employ less-skilled labour, and carry out work in

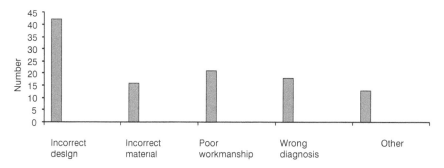

Figure 14.15 Causes of repair failure.

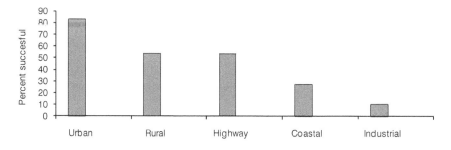

Figure 14.16 Influence of environment on repair performance.

unsuitable weather in order to avoid expensive delays. In many cases the causes of repair failure given by respondents could have been avoided if the financial pressures were less and good practice had been followed more closely.

There were no cases where the quality of materials was considered to be a cause of failure but there were many where it was considered that material selection had been incorrect. Repair materials should be compatible with the substrate particularly in relation to ductility. In many cases there has been undue emphasis on the strength of the repair material and inadequate attention to ductility.

The case histories have shown that there are benefits from the application of coatings on repairs as the success rate for patches was increased from 30 to 50 per cent and for crack injection from 70 to 80 per cent.

The performance of CP is encouraging as 60 per cent of the installations were reported to be wholly successful compared with 50 per cent for all repair methods. However, it is important for owners of structures to be aware that installations require regular inspection to ensure that they are in working order, and to be aware that they are susceptible to vandalism.

Abbreviations

AAR	Alkali–aggregate reaction
C Inj	Crack injection
C Inj+C	Crack injection plus coating
CP	Cathodic protection
Patch+C	Patch plus coating
X constr.	Faulty construction

References

Commission of Inquiry into the Collapse of a Portion of the de la Concorde Overpass (2007) *Report*. October 13–15, Québec: Bibliothèque et Archives nationales du Québec, Library and Archives. Canada. Gouvernement du Québec.

Cusson D and Mailvaganam N (1996). 'Durability of repair materials'. *Concrete International*, 1996, 18(3): 34–38.

Doyle N (1993). 'Beam failure starts US stadium check'. *New Civil Engineer*, 3 June, 4–5.

Emmons P (2006). 'Vision 2020: A strategic plan for improvements to the concrete repair industry'. Unpublished lecture, BRE Garston.

Matthews S, Sarkkinen M and Morlidge, J (2007). *Achieving Durable Repaired Concrete Structures: Adopting a Performance-Based Intervention Strategy*. IHS BRE Press, EP 77.

Oliver A (2000). 'Rusty tendons are likely culprit in US bridge collapse'. *New Civil Engineer*, 25 May, 5–7.

Radic J (1989). 'Bridge durability parameters'. *Symposium on the Durability of Structures*. Lisbon. IABSE.

Rowe G H, Freitag S A and Stewart P F (1984) *Concrete Quality in Bridges: New Zealand.* Central Laboratories Report 4 – 84/4.

Tilly G P (2002a). *Conservation of Bridges.* Spon, London.

Tilly G P (2002b). 'Performance and management of post-tensioned bridges'. *Proc ICE. Structures and Buildings*, 3–16.

Tilly G P and Jacobs, J (2007). *Concrete Repairs: Performance in Service and Current Practice.* IHS BRE Press, EP 77.

Van Begin C (1987). 'Durability of concrete bridges in Belgium.' Katherine and Bryant Mayer Conference. ACI Special Publ SP100, 1, 541–554.

15 Service-life modelling for chloride-induced corrosion

Rob B. Polder

15.1 Introduction

Since the mid-1990s, service-life modelling of concrete structures has become feasible and, subsequently, increasingly popular. Today many owners require service lives of 100 or even 200 years for important concrete infrastructure. Acknowledging previous work on modelling since the 1970s (Siemes et al., 1985), a major breakthrough was due to European Research project DuraCrete (DuraCrete, 2000; Siemes et al., 2000). DuraCrete successfully combined mathematical modelling of degradation processes with structural design philosophy, in particular reliability analysis. Since then, similar approaches and extensions have been worked out, additional field work has been done and particular subjects have been refined (Polder and Rooij, 2005; Gehlen, 2000; Li et al., 2008; FIB, 2006). Over the last five years, a simplified version has been developed in the Netherlands for chloride-induced corrosion (CUR, 2009). In principle, the approach is aimed at service-life design of new structures. However, modelling the remaining life of existing structures could be just as important for maintaining our infrastructure. Some work on degradation modelling was extended in that sense (Polder and Rooij, 2005; Rooij and Polder, 2005) but the subject is definitely underdeveloped. The main body of this chapter is devoted to describing the simplified model for chloride-induced corrosion related service-life design. However, in the final sections, possible application to existing structures is discussed.

15.2 Modelling for design of long service life

Until recently, design codes did not give quantified guidance for designing concrete structures for service life, in particular not for periods of 100 or more years. Another drawback of existing regulations is that no distinction is made between cement types (EN 1992-1-1, 2004). In particular, blending Portland cement with blast furnace slag, fly ash or silica fume has been found to substantially increase the resistance against chloride penetration

(Polder and Rooij, 2005; Polder, 1997; Bertolini et al., 2004; Bamforth and Chapman-Andrews, 1994; Rooij et al., 2007).

Design for 100 years or more requires some form of mathematical modelling that extrapolates degradation processes in concrete structures beyond our modern experience. Various models have been developed (Collepardi et al., 1972; Maage et al, 1996; DuraCrete, 2000; Siemes et al., 2000) and in the early 2000s significant experience was gained with the DuraCrete model in some major construction projects in the Netherlands. In 2003 a Dutch industry-wide committee, CUR VC81, set out to develop a practical probability-based guideline for service-life design of structural concrete with service lives up to 200 years, based on DuraCrete. Due to limited experience with this method, it was agreed that the requirements of the prevailing Dutch concrete standards should be taken as upper limits, which corresponded to international regulations (e.g. EN 206). This implied the usual maximum water-to-cement ratios and minimum cement contents depending on environmental class, with additional requirements to the minimum length of the curing period. Under these conditions chloride-induced rebar corrosion is likely to be the dominant mechanism determining the service life, whereas carbonation-induced corrosion can be ruled out safely. The Guideline was published in Dutch in 2009 (CUR, 2009) and is described in (Polder et al., 2010). It should be noted here that the Dutch climate is relatively mild and wet, resulting in moderate de-icing salt loads and limited carbonation in concrete with at least a reasonable quality. In other climates, the chloride load could be more severe (e.g. for marine conditions in warmer climates or for de-icing salt exposure in colder climates) or carbonation could be more important (e.g. for climates with longer dry periods). Carbonation modelling is discussed in (CEB, 1997), see also (FIB, 2006).

The basis for a probabilistic performance-based methodology for service-life design (SLD) was conceived in the 1980s by Siemes et al. (Siemes et al., 1985) and developed in detail in the 1990s in European research project DuraCrete (DuraCrete, 2000; Siemes et al., 2000). It follows structural limit state design philosophy by stating that the service life is the period in which the structure's *resistance* $R(t)$ can withstand the environmental *load* $S(t)$. $R(t)$ and $S(t)$ are time-dependent, statistically distributed variables. A particular (specified) *performance* is predicted with a predetermined maximum probability of failure at the end of the design life, as shown schematically in Figure 15.1. Scatter, variation and uncertainties, e.g. in material properties, geometries (concrete cover) and model parameters, are to be taken into account by probabilistic considerations.

The limit state is assumed to be initiation of reinforcement corrosion due to ingress of chloride ions. When the chloride content at the surface of the reinforcing bars exceeds the critical chloride threshold, the structure is considered to fail. The *load* is represented by the chloride content at the steel surface that increases with time due to chloride ions building

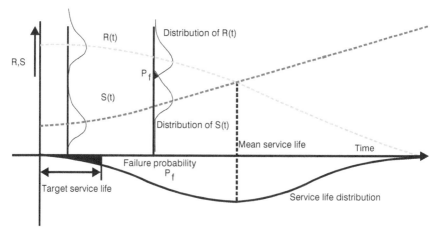

Figure 15.1 Schematic representation of probabilistic service-life design (Source: DuraCrete, 2000, Siemes et al., 2000).

up at the concrete surface and their subsequent transport into the cover concrete. The *resistance* is the critical corrosion initiating threshold for chloride at the steel surface. This latter property cannot reliably be influenced with present-day knowledge (Polder, 2009). As a practical approach, the critical chloride content is assumed constant in time. The concrete resistance to chloride penetration, expressed by its chloride diffusion coefficient, is taken as the model variable that, together with the chloride surface content, determines the load side of the service-life (limit state) equation. The chloride diffusion coefficient changes with time due to hydration, binding of ions and drying out, which complicates the modelling. It is emphasised that the *performance* is considered in terms of absence of corrosion initiation, which is not an ultimate limit state, because no direct danger for human lives is at stake. It is a maintenance limit state (MLS), because corrosion means the upcoming need to repair, which is an economic threat rather than a safety issue, corresponding to a serviceability limit state (SLS). Target probabilities of failure for SLS are usually in the range of up to 10 per cent.

Chloride transport modelling in DuraCrete was based on literature and exposure site data that mainly represented Portland cement (CEM I) concrete, with the exception of one set of exposure data (Bamforth and Chapman-Andrews, 1994). Additional information has since become available from systematic exposure (Tang and Andersen, 2000). More data on blast furnace slag cement (CEM III/B) concrete were provided by field work on six marine structures in the Netherlands. Chloride profiles were taken at ages of 20 to 40 years (Polder and Rooij, 2005; Rooij and Polder, 2005) from which modifications of model parameters were deduced. Additional data on CEM III/A and fly ash containing binders were obtained from participants in the CUR committee.

Consequently, the Guideline aims to cover service-life design of concrete structures in marine (exposure class XS) and de-icing salt environments (XD) for a wide range of binders. Chloride migration test results are used in a semi-probabilistic concept, which was condensed into a set of tables specifying maximum values for chloride diffusion coefficients depending on cover depth, cement type, exposure class and required service life.

15.3 A probability-based predictive model

The basic components of the present concept are a transport model, a chloride transport coefficient, several model parameters and a semi-probabilistic approach. For the description of chloride ion transport into concrete a modification of the error function solution to Fick's second law of diffusion is adopted. Using Fick's second law and the term diffusion coefficient does not exclude, however, that other mechanisms may contribute to the transport of chloride ions into concrete. For the present modelling, all non-diffusion contributions are assumed to be small and sufficiently dealt with by parameters in the model.

15.3.1 Chloride penetration (diffusion) coefficient

Since the diffusion model for chloride ingress in concrete was introduced by Collepardi et al. (Collepardi et al., 1972), several methods have been proposed for determining the resistance of concrete against chloride penetration. In the 1990s, two methods were standardised in the Nordic countries:

1 An immersion (pure diffusion) test, NT Build 443;
2 An accelerated (migration) test, NT Build 492, the rapid chloride migration test (RCM).

The immersion test may be seen as a realistic representation of the natural diffusion process. A drawback is that it requires seven weeks exposure of specimens and involves chloride analysis of many samples. The migration test involves a different transport mechanism (acceleration by an electrical field), but has a short execution time and is less laborious. In European research project Chlortest (Chlortest, 2005) both methods were studied and compared in a round-robin test. A good linear correlation was found between chloride diffusion coefficients from diffusion experiments and chloride migration coefficients from RCM tests. It was considered justified, therefore, to use the less time and labour consuming RCM test instead of the diffusion test (Breugel et al., 2008).

In the past few years, RCM testing has been applied to many concrete mixtures in association with service-life design of large infrastructural projects in the Netherlands. For the guideline, a total of 500 RCM-values from 153 different concrete compositions (precast and ready mix) were provided by participants. The influence of the mix composition on the RCM-value was analysed. Cement types used were mainly Portland cement (CEM I 32.5R, 52.5N, 52.5R), blast furnace slag cement (CEM III/A 52.5 R, CEM III/B 42.5 LH HS), mixtures of these two cements and binders comprising powder coal fly ash and Portland or slag cement. Binder contents ranged from 300 to 450 kg/m³ and water/binder ratios (w/b) from 0.33 to 0.65. The age at testing ranged from 28 days to 3 years, with most data at an age around 28 days.

It was assumed that the RCM-value depends to a large extent on the type of binder (cement type and reactive additions), w/b ratio and age. The data were first grouped with respect to binder type: Portland cement (CEM I); slag cement (CEM III/A or III/B, 50 – 76% slag); Portland and slag mixtures (25–38% slag); Portland cement with fly ash (21–30% fly ash).

Within these groups, data of similar age were aggregated; ages of 28 to 35 days were considered as a single group. The influence of the w/b ratio on D_{RCM} was then analysed for each particular binder group at an age around 28 days. All fly ash present in the mixture was considered as cementitious material, so w/b = water / (cement + fly ash). This analysis showed that the D_{RCM}-value is linearly related to the water/binder ratio:

$$D_{RCM}(28d) = A(w/b) + B \qquad (15.1)$$

with A and B constants for particular cement types, see Figure 15.2. This figure clearly shows that D_{RCM} and its w/b dependency strongly depend on binder type. For Portland cement D_{RCM} is strongly influenced by w/b. For cement with 50 to 76% slag this influence turned out to be less pronounced. The regression coefficients (A, B) found for different binders are in good agreement with those reported by Gehlen (2000). The data suggest that in the range of practical w/b ratios between 0.35 and 0.55 (and present-day concrete technology), minimum RCM-values apply that depend on binder type.

Even though execution of the RCM test is much faster than the diffusion test (NT Build 443), it is still labour intensive. For a quick impression of the resistance against chloride ingress the two-electrode method (TEM) for determining the electrical resistivity of concrete can also be used. In (Breugel et al., 2008) a good correlation was reported between RCM and inverse TEM results. Theoretical and empirical data support this correlation (Andrade et al., 1994; Polder, 1997). Particularly for production control the TEM test is suitable for a quick indication of the potential of a certain mixture to meet prescribed diffusion levels. If resistivity testing indicates non-compliance (i.e., a lower resistivity is measured than a predetermined minimum value that corresponds to the fixed maximum RCM value), cores can be taken and tested for RCM (Rooij et al., 2007).

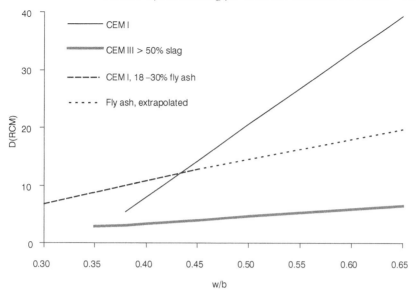

Figure 15.2 Correlation between w/b and D_{RCM} at about 28 days; all values* 10^{-12} m^2/s.

15.3.2 Modelling chloride ingress

In the DuraCrete model, the evolution of chloride profiles is approximated with:

$$C(x,t) = C_s - (C_s - C_i) \, erf \left[\frac{x}{\sqrt{\{4k\,D(t)\,t\}}} \right] \qquad (15.2)$$

where $C(x, t)$ is the chloride content (all chloride contents in this chapter are expressed in percentage by mass of binder) at depth x at time t, C_s is the surface chloride content, C_i the initial chloride content of the concrete, k is a correction factor, $D(t)$ is the apparent diffusion coefficient as a function of time (see below) and t is time. The surface chloride content was assumed to be independent of mix composition for reasons of simplicity: 3.0% for marine structures (Polder and Rooij, 2005; Rooij and Polder, 2005) and 1.5% for land-based structures (exposed to de-icing salts), based on data from (Gaal et al., 2003). The initial chloride content was taken equal to 0.1%.

The apparent diffusion coefficient $D(t)$ is multiplied by a correction factor k to obtain the chloride diffusivity of concrete in a real structure. This correction factor depends on binder type, environment and length of wet curing. Some of the deviations from pure diffusion behaviour are included in this parameter. The k-values were taken from DuraCrete or interpolated (Duracrete, 2000). The critical chloride content was taken to be equal to

0.6% by mass of cement, see (Polder, 2009; Gehlen, 2000; Gaal et al., 2003; Breit, 2001). This value applies only for concrete exposed to air, that is in environmental classes XS1, XS3, XD1, XD3. In XS2, concrete permanently submerged in sea water, the critical content is higher due to low availability of oxygen, except for structures that are submerged in sea water on one side and aerated on the other side (so-called hollow leg).

15.3.3 Time dependency of chloride penetration coefficient D(t)

The rate at which chloride ions penetrate into concrete decreases with time. This is due to hydration of the binder, which causes narrowing of capillary pores (especially in binders with slag or fly ash); and drying, which reduces the amount of liquid in the pores available for chloride transport. A decreasing rate of chloride penetration can be described with a time-dependent diffusion coefficient (Maage et al., 1996):

$$D(t) = D_0 \left(\frac{t_0}{t} \right)^n \tag{15.3}$$

where D_0 is the D_{RCM}-value at reference time t_0 (usually 28 days) and n is the ageing coefficient ($0 < n < 1$). The value of n for a particular binder in a particular environment depends on the rate of hydration and on the extent of drying. Ageing coefficients for different binders without the effect of drying were determined from D_{RCM}-values that had been wet (immersed) cured at 20°C for different periods from 28 days up to 3 years. In those experiments, slag and in particular fly ash binders showed higher n-values than Portland cement. This corresponds to results of chloride profiles from cores taken in the field (Gehlen, 2000). Structures in the field will dry out to some extent and hydration may occur more slowly than under water. In

Table 15.1 Ageing coefficients in equation (15.3) for different binders in two groups of environmental classes

Environmental classes	Underground, splash zone	Above ground, marine atmospheric
Type of binder	XD2, XS2, XS3	XD1, XD3, XS1
CEM I	0.40	0.60
CEM I + 25–50% slag, II/B-S; III/A, < 50% slag	0.45	0.65
CEM III, 50–80% slag	0.50	0.70
CEM I with 21–30 % fly ash or CEM II B/V	0.70	0.80
CEM V/A with c. 25% slag and 25% fly ash	0.60	0.70

DuraCrete, ageing coefficient values under field conditions were determined from profiles taken from structures and exposure tests, combined with *k*-values (equation (15.2)) for different environments and lengths of wet curing. Based on those data, the present analysis and additional work, in particular on blast furnace slag cement (Polder and Rooij, 2005; Rooij and Polder, 2005), *n*-values were chosen in two groups of environmental classes: very wet (XD2/XS3; also XS2) and moderately wet (XD1/XD3/XS1), see Table 15.1.

The time needed to reach the critical chloride content at a certain depth can now be calculated for any given mix (with a composition covered by the available database) in exposure classes XS and XD, using equations (15.1), (15.2) and (15.3) indicatively. For a specific service-life design calculation, D_{RCM} of a particular mix should be measured and the test result should be used for the calculation.

15.3.4 Reliability considerations and semi-probabilistic approach

For a given concrete cover depth, equation (15.2) can be used for calculating the time needed for the critical chloride content to reach the reinforcement. A matching diffusion coefficient for the intended concrete at reference time $D_{RCM,28}$ can be calculated. Such a calculation, however, is deterministic and yields a mean value. This means that the probability of corrosion initiation at that point in time is 50%. Consequently, the structure could have corrosion initiation over 50% of its surface! In practice such a high probability is unacceptable. An acceptable probability of failure for an MLS may be 10% for reinforcing steel (Fluge, 2001).

To obtain such a lower probability of failure (10%) either the cover depth can be increased or the maximum $D_{RCM,28}$ can be decreased. The latter option has serious disadvantages due to concrete technological limitations. If the former option is chosen, the required amount of additional cover can be calculated for each individual case using probabilistic calculations. This option is provided by the guideline, among others, by specifying mean values, standard deviations and distribution types for all parameters. However, another option given by the guideline is based on adding a fixed amount to the (deterministically determined) cover depth *as a safety margin*. This is a semi-probabilistic approach, comparable to using a safety factor for a materials property or a load. Calculations using TNO's probabilistic software Prob2B™ have shown that a safety margin of 20 mm results in a probability of failure of about 10%; a safety margin of 30 mm produces a probability of about 5%. Such probabilities are considered appropriate for corrosion initiation of reinforcing and prestressing steel, respectively.

Table 15.2 Maximum $D_{RCM,28}$ for various cover depths as a function of binder type and environmental class for a design service life of 100 years

| Mean cover [mm] | | Maximum value $D_{RCM,28}$ $[10^{-12}\ m^2/s]$ | | | | | | | |
| Reinforcing steel | Prestressing steel | CEM I | | CEM I+III 25–50% slag | | CEM III 50–80% slag | | CEM II/B-V CEM I + 20–30% fly ash | |
		XD1/2/3, XS1	XS2 XS3	XD1/2/3, XS1	XS2 XS3	XD1/2/3,XS1	XS2, XS3	XD1/2/3, XS1	XS2, XS3
35	45	3.0	1.5	2.0	1.0	**2.0**	1.0	6.5	5.5
40	50	**5.5**	2.0	**4.0**	1.5	**4.0**	1.5	**12**	10
45	55	**8.5**	3.5	**6.0**	2.5	**6.0**	2.5	**18**	15
50	60	**12**	5.0	**9.0**	3.5	**8.5**	3.5	**26**	22
55	65	17	7.0	12	5.0	*12*	5.0	*36*	30
60	70	22	9.0	16	6.5	*15*	6.5	*47*	39

Note: **Boldface** values are practically achievable by present-day concrete technology with currently used w/b ratios; italic values cannot be achieved by conventional concrete technology (low values) or are not recommended (high values) as they represent concrete with high w/b ratios.

15.4 Service-life design in practice – examples

Following the method described above, including a safety margin to the cover depth of 20 mm for reinforcing steel or 30 mm for prestressing steel, combinations of required cover depth and maximum D_{RCM} values (at reference time 28 days) were calculated for service lives of 80, 100 or 200 years. Wet curing was assumed to last for three days (XD) or seven days (XS). Values for 100 years are presented in Table 15.2. The following two examples illustrate using it for service-life design.

Example I concerns a reinforced concrete structure in XD1-3/XS1 environment. For the type of cement a CEM III/B with 70% slag was chosen. The required service life is 100 years. From Table 15.2 it can be seen that with a cover depth of 45 mm (reinforcing steel), a maximum $D_{RCM,28}$ is required of 6.0×10^{-12} m^2/s. With this cement and a *w/b* of 0.45, a D_{RCM}-value of 4.0×10^{-12} m^2/s can be obtained rather easily (Figure 15.2). Going back to Table 15.2 it can be seen that with a D_{RCM}-value of 4.0×10^{-12} m^2/s the cover depth could be reduced to 40 mm.

Example II concerns the same structure as Example I. The cover depth is 45 mm, but now CEM I is used. For CEM I and a cover depth of 45 mm, Table 15.2 gives a maximum $D_{RCM,28}$ of 8.5×10^{-12} m^2/s. Even though D_{RCM} is higher than in Example I, such a value might be hard to achieve with CEM I (see Figure 15.2). It would require quite a low *w/b*, probably below 0.4, which may cause workability problems. Increasing the cover to 50 mm will allow an increase of $D_{RCM,28}$ to 12×10^{-12} m^2/s, which can be readily achieved with a *w/b* of about 0.45.

Navigating through all possible options in the tables, the designer can find the economic optimum, while it can be demonstrated to the client that the required service life is achieved.

15.5 Application to existing structures

For existing structures, the situation is obviously quite different from that for design of a new structure. In a design situation, assumptions have to be made for material properties, cover depth and environmental load; the possibilities for obtaining information from testing in the intended environment are limited. For an existing structure, the material composition and cover depth cannot be changed. However, testing can reveal cover depths, material composition and other properties (diffusion coefficient, electrical resistivity) and the structure's response to the prevailing environmental load (chloride profiles, carbonation depths). This requires a detailed inspection including sampling and laboratory testing. It should be noted that considerable numbers of samples may be required to obtain a sufficiently representative picture, as scatter tends to be high (Polder and Rooij, 2005; Rooij and Polder, 2005). It is also important to realise that the critical chloride content for corrosion initiation may deviate from the model value given above, depending on concrete composition and external conditions. In principle it can be determined for the structure, but presently a procedure that has been proven in practice is not available. In the ideal case (excluding significant carbonation), a good picture is obtained of chloride penetration up to the date of the inspection from core sampling. Chloride penetration profiles should be fitted to a time-independent diffusion equation of the form:

$$C(x,t) = C_S \left(1 - erf \left(\frac{x}{2\sqrt{Dt}} \right) \right) \tag{15.4}$$

The chloride surface content C_S obtained in this way may be considered representative for the prevailing environment for a structure at least 10 to 20 years of age (Polder and Rooij, 2005; Rooij and Polder, 2005). Similarly, the accumulated (apparent) diffusion coefficient at the point in time of the inspection can be established. Both, by the way, will have the character of statistically distributed parameters. For predictive purposes, two possible approaches to the future development of the diffusion coefficient D exist. The more conservative approach is to use the best-fitting diffusion coefficient as a fixed value and extrapolate towards future penetration. A less conservative procedure would be to assume a decreasing diffusion coefficient using equation (15.3) and the appropriate exponential factor for the cement type used from Table 15.1. The difference between these two approaches may be small, however, in particular if corrosion is soon to be expected. In cases where some corrosion is already present, this method can be used to predict its spread to wider areas of the structure, which may be relevant for choosing repair options.

15.6 Concluding remarks

In many cases obtaining a long service life for civil engineering concrete structures (bridges, harbour quays, tunnels, parking garages) is predominantly a matter of postponing the onset of rebar corrosion due to chloride penetration. Noted exceptions are structures with 'exotic' preventative measures such as stainless steel reinforcement or cathodic prevention. The approach does apply in principle, however, to cases where coatings or hydrophobic treatment are used (although the model parameters should be modified to include their effects). The most important parameters in the presented model-based approach for concrete with ordinary steel reinforcement are:

- chloride load from sea water or de-icing salt environments
- chloride transport by diffusion
- time-dependent diffusion coefficients
- coefficients taking into account environmental, curing and materials (binder type) influences

- an assumed critical chloride content for corrosion initiation
- determining the rapid chloride migration coefficient of the intended concrete.

This chapter presents a probability-based design procedure for determining combinations of cover depth and 28-day chloride diffusion coefficients that are required to guarantee a specified service life for the limit state 'corrosion initiation due to chloride penetration'. Target probabilities of failure are 10% and 5%, respectively, for reinforcing steel and prestressing steel. Based on a semi-probabilistic simplification, the required combinations of cement type, cover depth and diffusion coefficient are brought together in design tables. The tables give limiting values for chloride diffusion coefficients obtained with the RCM test for service lives of 80 to 200 years in marine (XS) or de-icing salt (XD) environments. From analysis of a large number of test results, the dependency of the D_{RCM}-value on w/b and cement type was determined and an indication was obtained in which values are practically possible using present-day concrete technology. Distinguishing cement types is considered important. Even the most recent international regulations do not seem to make such distinctions (EN1992-1-1, 2004).

Similar tables as proposed here have recently been presented by Li et al. (2008). In their tables, however, the compressive strength is still considered one of the durability parameters. Instead of strength, here an explicit transport parameter is chosen, i.e. the RCM-value, to indicate the concrete's susceptibility to chloride ingress. A similar probability-based approach to various degradation mechanisms has been presented by *fib* (FIB, 2006) using a slightly different model for chloride-induced corrosion.

With this Guideline (CUR, 2009), the Dutch concrete industry now has rules for practical service-life design for structural concrete. All parties involved have agreed to collect their experience using the Guideline, with the intention to evaluate it and if necessary, to improve it in the near future. At the same time, however, it was realised that many items used in the calculations still contain large uncertainties. Further research should contribute to reducing them. Finally, the approach taken may contain elements that are useful for determining the remaining life of existing structures.

Acknowledgement

This chapter is based on ideas and results from the DuraCrete consortium, researchers at TNO, INTRON and TUDelft and CUR-committees B82 and VC81. The financial and in kind support of Rijkswaterstaat and other participating organisations are gratefully acknowledged. Any errors are the author's.

References

Andrade, C., Sanjuan, M.A., Recuero, A. and Rio, O., 1994, 'Calculation of chloride diffusivity in concrete from migration experiments, in non steady-state conditions', *Cement and Concrete Research*, **24** no.7, 1214–1228.

Bamforth, P.B. and Chapman-Andrews, J., 1994, 'Long-term performance of RC elements under UK coastal conditions', in R.N. Swamy (ed) *Proc. Int. Conf. on Corrosion and Corrosion Protection of Steel in Concrete*, Sheffield Academic Press, 24–29 July, 139–156.

Bertolini, L., Elsener, B., Pedeferri, P. and Polder, R.B., 2004, *Corrosion of Steel in Concrete: Prevention, Diagnosis, Repair*, Wiley-VCH Verlag. KGaA, Weinheim.

Breit W., 2001, 'Critical corrosion inducing chloride content – State of the art and new investigation results', *Betontechnische Berichte 1998–2000*, Verein Deutscher Zementwerke e.V., Verlag Bau+Technik, Düsseldorf, 145–167.

Breugel, K. van, Polder, R.B. and Wegen, G.J.L. van der, 2008 'Service life design of structural concrete – The approach in The Netherlands', *Proc. Structural Faults and Repair*, Edinburgh.

CEB, 1997, *New Approach to Service Life Design: An Example for Carbonation Induced Corrosion*, CEB Bulletin 238, Lausanne.

Chlortest, 2005, Workshop: Resistance of concrete to chloride ingress – from laboratory test to in-field performance, December 1–2, University of Alicante (2005), see www.chlortest.org.

Collepardi, M., Marcialis, A. and Turriziani, R., 1972, 'Penetration of Chloride Ions into Cement Pastes and Concretes', *J.Am.Cer.Soc*, **55**, 534–535.

CUR, 2009, 'Durability of structural concrete with regard to chloride induced reinforcement corrosion – Guideline for formulating performance requirements', downloadable via http://www.curbouweninfra.nl/ CUR Gouda (in Dutch).

DuraCrete, 2000, *DuraCrete – Probabilistic Performance based Durability Design of Concrete Structures*, DuraCrete Final Technical Report R17, Document BE95-1347/R17, The European Union – Brite EuRam III, CUR, Gouda.

EN 1992-1-1, EuroCode 2, 1994.

fib, 2006, *Model Code for Service Life Design*, Bulletin 34, Model Code, fib.

Fluge, F., 2001, 'Marine chlorides – A probabilistic approach to derive provisions for EN 206-1', Third DuraNet Workshop on Service Life Design of Concrete Structures, From Theory to Standardisation, Tromsø, June 10–12.

Gaal, G.C.M, Polder, R.B., Walraven, J.C. and Veen, C. van der, 2003, 'Critical chloride content – state of the art', in M.C. Forde (ed), *Proc. 10th International Conference Structural Faults and Repair*, Engineering Technics Press Edinburgh.

Gehlen, C., 2000, *Probabilistische Lebensdauerbemessung von Stahlbetonbauwerken*, Deutscher Ausschuss für Stahlbeton, Berlin.

Li, K., Chen, Z. and Lian, H., 2008, 'Concepts and requirements of durability design for concrete structures: an extensive review of CCES01'. *Materials & Structures* **41**, 717–731.

Maage, M., Helland, S., Poulsen, E., Vennesland, O. and Carlsen, J.E., 1996, 'Service life prediction of existing concrete structures exposed to marine environment', *ACI Materials Journal* 93 (6), 602–608.

Polder, R.B., 1997, 'Chloride diffusion and resistivity testing of five concrete mixes for marine environment', in L.-O. Nilsson and P. Ollivier (eds) *Proc. RILEM International*

Workshop on Chloride Penetration into Concrete, St-Remy-les-Chevreuses, October 15–18, RILEM, 225–233.

Polder, R.B., 2009, 'Critical chloride content for reinforced concrete and its relationship to concrete resistivity', *Materials and Corrosion* **60** (8), 623–630.

Polder, R.B. and Rooij, M.R. de, 2005, 'Durability of marine concrete structures – field investigations and modelling' *HERON* **50** (3), 133–154.

Polder, R.B., Wegen, G. van der, and Breugel, K. van, 2010, 'Guideline for service life design of structural concrete with regard to chloride induced corrosion – the approach in The Netherlands', Second International Conference on Service Life Design, Delft, October, submitted.

Rooij, M.R. de and Polder, R.B., 2005, *Durability of Marine Concrete Structures*, CUR report 215, CUR, Gouda, (in Dutch).

Rooij, M. de, Polder, R.B. and Oosten, H. van, 2007, 'Validation of durability of cast insitu concrete of the Groene Hart railway tunnel', *HERON* **52** (4), 225–238.

Siemes, T., Vrouwenvelder, T. and Beukel, A. van den, 1985, 'Durability of buildings: a reliability analysis', *HERON* **30** (3), 2–48.

Siemes, T., Schiessl, P. and Rostam, S., 2000, 'Future developments of service life design of concrete structures on the basis of DuraCrete', in D. Naus (ed) *Service Life Prediction and Ageing Management of Concrete Structures*, RILEM, 167–176.

Tang, L. and Andersen, A., 2000, 'Chloride ingress data from five years field exposure in a Swedish marine environment', in C. Andrade, J. Kropp (eds) *Proc. 2nd International RILEM Workshop Testing and Modelling the Chloride Ingress into Concrete*, PRO 19, RILEM Publications, 105–120.

16 Case studies in the use of FRPs

Paul Russell

16.1 Introduction

FRP products (glass, carbon fibre and aramid) were first used to reinforce concrete structures in the 1950s. During the next two decades, the quality of the FRP materials improved considerably, the manufacturing methods became more automated and material costs decreased. The use of these materials for external reinforcement of concrete bridge structures started in the 1980s, first as a substitute for steel plate bonding, which had been a proven method for many years previously, and then as a substitute for steel confinement shells for bridge columns.

The technology for external retrofitting was developed primarily in Japan (wet lay-up sheet wrapping) and Europe (pre-cured laminate bonding). Today there are more than 1000 concrete slab/steel girder bridges in Japan that have been strengthened with wet lay-up sheet bonding to the slabs. Also, many thousands of bridge columns have been seismically upgraded with the same materials. Ongoing development of cost-effective production techniques for FRP composites has progressed to the level that they are now readily accepted in the construction industry. Owing to published design guidelines similar to the Concrete Society TR55 (Concrete Society, 2004) and 57 (Concrete Society, 2003) and CIRIA-C595 (CIRIA, 2004), engineers now have the confidence to design strengthening schemes that are a benefit to their clients and the users of the structures.

Reduced material cost, together with labour savings inherent with their low weight and comparably simpler installation, relatively unlimited material length availability, and immunity to corrosion, make FRP materials an attractive solution for post-strengthening, repair, seismic retrofit, and infrastructure security. Another advantage of the use of FRP materials is that in many cases the structures can be returned to their owners a lot more quickly than when using traditional materials such as steel and concrete. Savings in lane rentals (highways) and line possessions (rail and track) are a bonus to those people who use the structures, relieving traffic congestion or delays on the rail infrastructure.

The principles behind externally bonding FRP plates or wraps to concrete structures are very similar to the principles used in application of bonded steel plates. In general, the member's flexural, shear, or axial strength is increased or better mobilised by the external application of high tensile strength materials.

Reasons for applying FRP systems as an external reinforcement to bridge structures include:

- Capacity upgrade due to a change in use of a structure
- Passive confinement to improve seismic resistance
- Crack control
- Impact resistance
- Strengthening around new openings in slabs.

Reasons for applying FRP systems as an external reinforcement to other structures would include those mentioned above as well as:

- Blast mitigation
- Increasing in the confinement of the concrete members so as to add extra floors to structures
- Increasing capacity of members including steel and concrete.

FRP composite systems have been applied to many structural elements including beams, columns (including steel and cast iron), slabs and walls. They have also successfully been used in many special applications such as chimneys, pipes and tanks.

16.2 FRP systems

FRP system forms can be categorised based on how they are delivered to the site and installed. External FRP reinforcing systems come in a variety of forms including

1   Wet lay-up systems
2   Pre-cured systems (pultruded and pre-preg)
3   Near-surface mounted systems (NSM).

16.2.1 Overview of wet lay-up systems

Wet lay-up FRP systems consist of dry unidirectional or multidirectional fibre sheets or fabrics that are impregnated on-site with a saturating resin. The saturating resin is used to provide a binding matrix for the fibre and bond the sheets to the concrete surface. Wet lay-up systems are saturated with resin and cured in place and in this sense are analogous to cast-in-place concrete. Three common types of wet lay-up systems are:

- Dry unidirectional fibre sheets with the fibre running predominantly in one planar (0 axis) direction
- Dry multidirectional fibre sheets or fabrics with fibres oriented in at least two planar directions
- Dry fibre tows that are wound or otherwise mechanically applied to the concrete surface. The dry fibre tows are impregnated with resin during the winding operation.

16.2.2 Overview of pre-cured (pultruded and pre-preg) systems

Pre-cured FRP systems consist of a wide variety of composite shapes and profiles, manufactured in the system supplier's factory and transported to the job site. Typically, an adhesive is used to bond the pre-cured flat sheets, rods or shapes to the concrete surface or inserted into slots cut into the wall. The adhesive used to bond the pre-cured system to the concrete surface must be specified by the system manufacturer. Adhesive selection is critical in that the adhesive provides for the proper transfer of load between the surface of the concrete and the cured reinforcement. Common types of pre-cured systems are:

- Pre-cured unidirectional laminate sheets in the form of large flat plate stock or as pultruded laminates coiled on a roll.
- Pre-cured multidirectional grids and mesh coiled on a roll or pre-cut in sheet form. These grids are easily installed on the job site using tin-snips or suitable cutters.
- Pre-cured shells in the form of shell segments are fitted around columns or other structural elements and then bolted together and finally grouted to form a composite section, to provide seismic confinement or strengthening.

16.3 Strengthening techniques

There are three ways in which we generally can upgrade the structure, using concrete, steel or fibre-reinforced plastics (FRPs).

Figure 16.3 illustrates not only the speed but also the weight savings made by using composite materials. As can be seen from the example, the FRP alternative is far quicker to install, using less labour or man-hours and cures far quicker. An obvious saving using this method would be on traffic lane rental or on railway possessions.

Typical costs of rail possessions or lane rentals in the UK in 2006 were:

- Rail – Gerard's Cross tunnel, c. £1m per day
- Road – £12 per vehicle per hour; M25, 30 August lorry fire, 40,000 vehicles, 2 hours: £1m

Figure 16.1 Composite pre-formed shell made from Kevlar and epoxy resin.

Figure 16.2 Composite shell offered up to the RC column. The shell is then bolted together and grouted between the column and shell with a cementitous grout.

An alternative could be to demolish the structure and start again.

The cost of demolition for buildings is an impossible calculation: down time, re-building, cost of new materials etc. If we were to look at the environmental impact issue then we have the following statistics:

Simply supported beam; 35% upgrade in live load

Bonded Steel Plate

♦ 6mm bolted plate
♦ 110kg dead load
♦ 6 men to lift and place

Member Enlargement

♦ 2 #8 rebar,
♦ 100mm. grout
♦ 1110kg dead load
♦ Formed and cured

FRP Sheet

♦ 1 layer resin bonded
♦ 1.5kg. dead load
♦ Placed by hand

Figure 16.3 Methods of strengthening beams with FRP.

- 34% of the UK's waste is derived from construction and demolition activities.
- The USA has 594,000 bridges – 28% of them are structurally deficient or obsolete.

As previously mentioned, a good example of strengthening using composites would be a highway structure, where vehicles from the continent are delivering goods to the UK, and the strength and weight limits of our bridge structures are lower than those in parts of Europe. Over the last 15 years or so, local agencies have undergone a programme of strengthening the UK's bridge structures to meet the demands of European transport legalisation.

The technology and application techniques used to strength these structures have improved dramatically over those 15 years. The UK was using steel plates and larger sections of concrete to strengthen such structures. Today these methods are rarely used, favouring advanced lightweight composite materials such as carbon fibre, glass fibre and aramid fibres. The use of FRPs has progressed from using simple pultruded plates and wraps to the use of ultra high modulus systems and unique profiles.

The strength of the fibres that are in use today can be up to 10 times that of mild steel, while possessing a similar stiffness. For specialist applications, where stiffness is a concern, such as cast iron structures, fibres up to four times stiffer than steel are now available.

This chapter will examine some of the more intricate and interesting applications, where the engineers involved have designed and promoted the use of these very exciting materials. Where perhaps in the past they would have had to design using steel plates or concrete, both materials being very labour intensive and very awkward to install, nowadays FRP wrapping is routinely specified.

16.4 Harbour City multi-storey car park

In the majority of inner city areas, car parking poses a major problem to those people who insist on using their own transport. Recent Acts of Parliament have tried to ease the burden on both our roads and cities. Car parking spaces are often very hard to find, as well as being very tight in their size.

In some inner city areas, the possibility of building new car parks is not viable due to the lack of space available. If the present car parks were to be demolished then that would cause an even greater problem whilst the new car park was being erected. What then are the alternatives to this ever-growing problem?

People could try leaving their cars at home; this has been tried by many, but never seems to last. The solution then is to extend the existing parking structures by adding more floors, hence giving more spaces to those who need to park.

Harbour City Car Park is situated at Salford Quays, close to the new Lowry Centre. Office blocks and retail facilities are being erected all the time in this area causing a major headache to car drivers. The car park was built in 1991, designed by Allott and Lomax, now a part of Jacobs Engineering (UK) Ltd. The building is an in-situ concrete structure, with brick cladding panels. The car park has won numerous awards for its brickwork cladding panels.

It was decided that an extra 300 spaces were required. To do this an extra four floors were needed, these being split into two levels. What were the solutions available to this structure? Demolition and rebuilding was quickly dismissed as this would have caused major problems during the reconstruction of the car park, as there was simply nowhere else to park during the works. Traditional methods such as extra reinforcement and concrete to the columns were one option, the other being steel plates or hoops.

It didn't take too long to realise that, if these methods were adopted, then valuable car parking spaces would be lost due to the extra dimensions added to the columns. Another consideration would be the weight factor, which had to be kept to a minimum. The only effective solution left was to strengthen the columns by confining them using FRPs. It was decided that a Kevlar[1] fibre-based system, MBrace Kevlar, would be the best solution.

A very tight construction programme had to be met; all the strengthening work had to be completed in 6 weeks, as well as the car park remaining open at all times.

---

1 Kevlar® is a registered trademark of DuPont.

To prepare the concrete, the corners of the columns were ground to a 15–20 mm radius, and then they were water jetted to remove any laitance. All of this work was carried out at night when the car park was empty. Cutting of the fibres was carried out using mechanical shears to quicken the application. All of the fabrics, where possible, were cut off site, at the offices of the specialist contractor. As the columns were of similar size, this method was found to be quicker.

Mixing of the resin at the preparation area made the job simpler. A resin application area was made at the job site; two lengths of Kevlar® could then be saturated, rolled up and moved to the prepared columns.

The columns were then covered with the Kevlar in the hoop direction. A minimum of two layers was specified with a minimum overlap of 200 mm. The coating system was then applied the following day and the car park spaces were passed back to the client.

Eight years after the car park was strengthened and the extra floors were added to the structure, there are no signs of distress to the structure; in fact more clients are looking at this method as a solution to the shortage of car parking in our inner cities.

16.5 West Burton Power Station

The cooling towers at West Burton Power Station were built in the 1960s. It has eight reinforced concrete cooling towers, each 118 metres high with a 46.3 metre internal throat diameter and a distinctive hyperbolic profile. The original tower shells are only 125 mm thick, making them thinner than an eggshell by scale.

Since the 1970s, when a number of high-profile cooling tower collapses such as Ferrybridge occurred, the issue of strengthening existing cooling towers has had to be addressed. Older cooling towers built to a CEGB standard design were under-designed and failed to take properly into account the effects of wind loading. Subsequent collapses led to a nationwide programme of inspection to prioritise those towers which required strengthening, and strengthening works throughout the country ensued.

Typically, the technique used to strengthen cooling towers, according to the CEGB design, involved building a 125 mm thick mantle – an outer skin of reinforced concrete – on top of the existing shell. While 125 mm was the standard thickness according to the design, mantles, however, were usually 75 to 100 mm thick. In the 1970s a 'gunite or shotcrete' (sprayed concrete) shroud was applied to the surface to help strengthen the shell, especially against wind loading.

After a survey by Jacobs Babtie, it was noted that the cooling tower was still continuing to crack and deform. The engineers came up with a solution to this problem by tying into the cooling tower a series of concrete ring beams to provide the required circumferential stiffness. Was this enough though to ensure that the tower would regain the anticipated strength and

Figure 16.4 Protection to the MBrace Kevlar using Masterseal 525 flexible cementitious coating.

Figure 16.5 Extra floors being added to the structure after Kevlar application.

stiffness? Jacobs Babtie addressed the one aspect that if left unattended could cause a problem. The relatively high stiffness that was now introduced into the structure would attract load and cause peak stresses under wind loading. As the beams were tied into the shell, they could cause cracking to the inside of the tower.

A lightweight, strong reinforcing material, which could be applied internally to the surface of the tower, was needed to counteract these stresses.

At this stage, construction chemical manufacturers and supplier BASF CC Ltd advised the engineer on the materials available. Both carbon fibre and aramid fibres were offered to the engineer. The material would be applied from cradles to the inside of the tower and so it would need to be robust enough to withstand application and future inspection, as the cradle wires which guide the cradles along the contours of the internal face had the potential to rub and cause mechanical damage.

Owing to the characteristics of carbon fibre, mainly its brittle nature, it was decided that this option would not be viable for this project. Kevlar® is an aramid fibre manufactured by DuPont, and has been used for more than 25 years in a wide variety of applications ranging from aerospace composites and body armour to protective apparel. Using MBrace Kevlar represented the obvious solution because of its low weight, high strength, and high stiffness and damage tolerance. The fabric drapeability ensured that it would be easy to apply, and would conform to the contours of the cooling tower wall.

The West Burton contract was the first application of MBrace Kevlar brand fabric to a cooling tower anywhere in the world, and the first major use of such fabric on a structure in Europe.

Before specialist subcontractor Balvac applied the Kevlar, several different kinds of surface preparation were tried, from grit blasting, needle gunning and concrete planing to keying the surface with a rotary wire brush. Concrete planing with localised needle gunning where required (for example, in existing depressions which planing could not reach), proved to be the most successful method. Not only was it quicker than needle gunning on its own, but it also avoided the disadvantages of grit blasting (dust, which would then have to be removed).

In all cases, pull-off tests on the Kevlar after application to the prepared surface led to the failure of the concrete, showing that the bond strength was greater than the tensile strength of the concrete.

After surface preparation, the sheets of Kevlar fibre were encapsulated in the two-part epoxy resin and applied to the wall of the tower using the wet lay-up technique.

As well as using circumferential bands of Kevlar to counteract the stresses caused by the circular strengthening beams, vertical strips were placed between the relatively stiff rings to prevent cracking. These were applied in deformed areas of the shell where analysis identified high stresses developing under wind loading. Because the Kevlar was to be protected with a flexible coating towards the end of the project, it was imperative that a good bond was achieved between the coating and the fabric.

Various methods are available to achieve this: simply scattering a suitable fine aggregate onto the wet resin to form a mechanical key, or alternatively using peel ply. Peel ply is a relatively inexpensive material that is placed onto the wet resin prior to curing. When the contractor is ready to apply the coating system he simply removes the peel ply. The advantage of using the peel ply in this situation is that it keeps the area clean; there is no need to wash the surface down if it has been left for a number of weeks. The mechanical key is already formed by the peel ply being pushed into the wet resin.

As already mentioned the Kevlar was coated with a cementitious flexible coating. The reason for this, apart from aesthetic reasons, was the protection it afforded to the installation as well as the structure. All the resins used in the wet lay-up system were based on 100% solid epoxy components. Epoxy resins are, however, susceptible to eventual degradation if not protected.

The advantages of using Kevlar on this particular project were: speed of application; toughness and durability – Kevlar is very resilient to abrasion whereas carbon is a very brittle material that would wear very quickly in this environment; finally, the drapeability of the material in the way it adapted itself to the shape of the structure.

The structure has now been in operation for several years and there have been no problems reported as to the functionality of the materials.

16.6 Rhuddlan Castle, North Wales

This is a good example of where newer technology has been used. Rhuddlan Castle is an ancient listed structure (Figure 16.6). The requirements for this project were for the stabilisation and stitching of the masonry back together. Owing to the strengths required by the designer, a high modulus carbon fibre bar was used. To help with the adhesion properties of the carbon fibre bar, it was coated at the time of manufacture with a peel ply finish, which, when removed prior to application, gave a very good mechanical bond. The mortar beds were raked out and the carbon bar inserted into the formed channels.

16.7 Trenchard Street multi-storey car park, Bristol

The adhesive used to fix the bars in place was based on 100% solid epoxy resin; although this resin is more of a putty-like consistency than the resins used in the wet lay-up systems such as the projects at West Burton Power Station and Harbour City MSCP.

BASF Construction Chemicals have had similar success in Europe using carbon bar. Examples of NSM (near-surface mounted) bars are now commonplace. This application tends to be in areas where seismic problems

Figure 16.6 Rhuddlan Castle.

are common or where a lower amount of cover is required on pre-cast elements.

Pre-cured NSM rod/shape systems can generally be used as an alternative for reinforcing concrete and masonry structures similar to surface laminates. NSM rods/shapes provide a more discrete solution to strengthening structures in that they generally are inserted into the masonry or concrete structure via slot or saw cuts. Generally, shapes can vary in size depending on application but typically are provided in round and rectangular cross-sections. The shapes are manufactured in the system supplier's facility and shipped to the job site. The shapes generally feature a surface treatment to facilitate bond between the FRP and adhesive or grout. An epoxy adhesive or cementitious grout is used to bond the pre-cured rods in the groove cut into the surface. Adhesive selection is critical in that the adhesive provides for the proper transfer of load between the wall and the cured reinforcement. A cosmetic surface can then be added to completely hide the strengthening system. Since the products are embedded into the substrate and bonded on three sides of the FRP shape, development lengths for NSM strengthening may be shorter. NSM rods/shapes may also be anchored into adjacent members and the opportunity of upgrading elements in their negative-moment region is opened up, as the FRP shape is not exposed to potential mechanical damage typical of floor or deck systems. FRP rods/shapes using the NSM technique

Figure 16.7 Application of the bonding resin injected into the stone.

do not require extensive surface preparation and installation time may be less than for other systems.

Installation/application

After assessment of the condition of the existing structure and design by a competent professional, installation of the NSM FRP strengthening is performed according to the following general guide:

- Cut groove – Using a diamond blade saw or grinder, a groove 1.5 times the bar diameter (in the case of a rectangular FRP shape, 1.5 times the depth and 3 times the thickness) is cut as prescribed. The use of two diamond blades on the saw arbor may be necessary.

Pre-cured shapes for near surface-mounted (NSM) FRP

- Prepare groove – The groove is masked with masking tape or similar product to prevent excess adhesive from marring the surface. The groove is thoroughly cleaned using a vacuum and/or compressed air.
- Apply adhesive – Structural adhesive gel or grout is filled in the groove. Care should be taken to avoid entrapped air voids.
- Place FRP rod/shape into groove – After the adhesive has been applied into the groove, the rod is placed and pressed into the groove to ensure proper location of the rod/shape.
- Finish – After the FRP rod/shape is seated into the groove, the adhesive is smoothed and any additional adhesive is added. The location is then cleaned up and the masking removed.

Figure 16.8 Placing carbon fibre bar into prepared slot.

Figure 16.9 Application of the epoxy adhesive to fix the carbon fibre bar.

16.8 QAFCO, Qatar, UAE

A good example of carbon fibre laminates (pre-cured systems) has been their use to strengthen a Prill tower in the Middle East; 85 rings of MBrace carbon fibre laminate were used in this application. The original design looked at using steel plates or cast-in-situ concrete beams. After careful analysis of cost and down time it was decided that carbon fibre was the only viable solution for this project.

The problems associated with this project were high temperature, both air and operating temperature, as well as difficult access.

A very tight programme was required, with the original shut-down being only 20 days. The contract for the application of the carbon fibre laminates was carried out in just 17 days, which is a credit to the supervising engineers, the contractor as well as the material supplier.

The three-day saving in the application stage gave an overall cost saving in extra production time for the Prill manufacture valued at over £1 million.

The specification for the laminates called for a peel ply to be used; this helped in the mechanical adhesion of the laminates. The resin used was developed for this project to take into consideration the temperature at the time of the project. The choice of resin was very important for this project; the resin and the laminate had to withstand a temperature of 60°C. Tests carried out by Oxford Brookes University proved that the laminate and resin would withstand the high temperature on this contract. The tests concluded that the resin actually increased its glass transition temperature (T_g).

Figure 16.10 QAFCO Prill Tower, Qatar.

The adhesive was developed originally for the contract carried out in Qatar for QAFCO where the parameters for the adhesive were that it had to have a T_g of > 65°C. The obvious advantages of using carbon fibre laminates instead of steel plates were that no holes had to be drilled into the concrete substrate to keep the plates in position. The other advantage was the time element, as well as a reduced labour requirement on the job site.

16.9 Tay Bridge, Dundee

The Tay Bridge is again a very good example of where MBrace Kevlar has been used to strengthen the columns both for confinement and for shear to the north approach road. The columns in this case were wrapped with Kevlar.

The columns are 6 m high, and the criterion was that the columns had to be able to carry 40 tonne vehicles. The designers, Babtie, decided on using a thicker aramid sheet to save time on the contract. MBrace Kevlar AK90 was chosen for this.

The shape of the columns dictated the method of application. As the columns were oval shaped as well as tapering, the material was applied in two halves with an overlap of 200 mm.

16.10 St Michael's road bridge, NW England

St Michael's Bridge is a good example of how the speed of application was a major influence in its use. Earlier in this chapter we looked at the cost of possessions on the West Coast main line. St Michael's Bridge is not on this main line, but is still on the main route from Manchester to Liverpool. This is the first overbridge to be strengthened in possession for Network Rail.

Details of the bridge are as follows:

- RC train carrying (underline) bridge
- Original bridge metallic beams and cast iron jack arches – mid-19th century
- Deck replaced 1959
- 4 pre-cast RC concrete deck units carry the track
- 2 pre-cast RC parapet units
- Deck under strength in flexure due to increased live load.

The criteria was that the bridge had to carry 140 mph Virgin trains, so the engineer and client looked at the various options available to them. The options were to use steel plates for equivalent strengthening, with the following requirements:

- 270 MPa tensile strength, 210 GPa modulus
- 8–10 mm minimum thick plates

Figure 16.11 Carbon fibre laminates with temporary supports.

- Lines of bolts every 150 mm
- Significant amount of drilling of the soffit for the bolts
- At least 12–14 possessions to carry out the work
- Longer duration on site than CFP meant a cost of some £400,000
- Extensive disruption to operations.

The Engineers and Client concluded that carbon fibre plates were the most appropriate solution because:

- They are a lightweight material
- Fast, simple installation
- Low risk during installation
- Minimal plant requirements
- Opportunity to carry out the works in normal rules of the route possessions
- Minimal increase in overall depth of the slab (headroom maintained).

Figure 16.12 St Michael's Bridge.

Figure 16.13 Carbon plates applied to the soffit of the bridge.

CFP also meant a long-term durability due to excellent properties (fatigue, creep, corrosion) of the carbon fibre materials and furthermore the whole project cost less than £90,000. The work was carried out in a total of four night-time possessions.

16.11 Conclusions

Carbon fibre sheets, plates and rods can be used in a wide variety of strengthening situations. Their light weight and immense strength make them easy to install and quite able to deal with the loads imposed on them.

They represent an effective, low-cost solution to the strengthening and repair of structures.

References

CIRIA (2004) *Strengthening Concrete Structures with Fibre Composite Materials: Acceptance, Inspection And Monitoring*, Report C595, London: CIRIA.

Concrete Society (2003) *Strengthening Concrete Structures with Fibre Composite Materials: Acceptance, Inspection and Monitoring*, TR57, London: The Concrete Society

Darby, A., Ibell, T. and Clarke, J. (2004) *Design Guidance for Strengthening Concrete Structures Using Fibre Composite Materials*, TR55, London: The Concrete Society.

17 Coatings for concrete

Shaun A. Hurley

17.1 Introduction

Coatings, normally in the form of proprietary products, are commonly used on mass or reinforced concrete to:

- Prevent premature deterioration
- Limit or control ingress of gases or aggressive chemicals
- Enhance appearance
- Modify other surface properties; for example, water repellency, slip resistance, colour or impact and abrasion resistance.

Products are available for internal, external, underwater and trafficked surfaces.

The following related products are outside the scope of this discussion:

- Thick renders, floor toppings and gunite
- Waterproofing membranes for car parks/bridges
- Preformed sheet membranes for waterproofing
- Protective tapes
- Curing membranes
- Fire protection systems
- Coatings associated with electrochemical techniques that halt or prevent reinforcement corrosion (cathodic protection, desalination or realkalisation)
- Coatings for steel reinforcement and active corrosion inhibitors.

17.2 Why use coatings?

Uncoated concrete provides a long service life in many environments. In aggressive conditions requiring additional surface protection, concrete can still remain an attractive construction material due to its versatility and relatively low cost. For some applications, protection against various forms of deterioration and/or ingress/transmission may be essential; for others, it may be optional, giving increased assurance of satisfactory durability. However, coatings should not be viewed as a basis for reducing cover or for inadequate mix design, placement and curing (although they are sometimes used as a remedial measure following these actions). The need for enhancement/ change of appearance and/or the modification of other surface properties may also vary from being essential to optional.

Wherever surface coating/treatment is optional, increased initial costs, and the inevitable maintenance costs, must be balanced against the projected in-service benefits.

A summary of the common reasons for applying coatings/surface treatments to concrete is given in Table 17.1. Although these applications are well proven, specific local conditions can affect performance significantly. Consequently, particular requirements should always be discussed with suppliers.

More specifically in relation to the protection and repair of concrete structures, general principles for the use of products and systems – including coatings and related treatments – are set out in the European Standard BS EN1504-9: 2008 (BSI, 2008).

These principles, and the related methods that are relevant here, are summarised in Table 17.2. As shown, there are eleven principles, six of which are concerned primarily with defects caused by mechanical, chemical or physical actions on the concrete itself, and five of which are specifically concerned with reinforcement corrosion. A related European Standard BS EN1504-2:2004 (BSI, 2004) deals in detail with the performance properties of these products and systems.

When applying surface protection to concrete, BS EN 1504-2:2004 suggests that products can be assigned to three main categories:

- Hydrophobic impregnation (H) – these materials penetrate the concrete and leave a water-repellent, molecular lining on the surface of the pores. They encourage the concrete surface to shed water but do not prevent water ingress under significant pressure as the pores essentially remain open.
- Impregnation (I) – these materials impregnate the concrete and block the pores.
- Coatings (C) – coating systems are those that adhere to the outer surface of the concrete.

These in turn relate to the repair principles in BS EN 1504 Part 9:

- Principle 1: Protection against ingress (PT)
- Principle 2: Moisture control (MC)
- Principle 5: Physical resistance/surface improvement (PR)
- Principle 6: Resistance to chemicals (RC)
- Principle 8: Increasing resistivity by limiting moisture content (IR)

Table 17.1 Common reasons for using surface coatings/treatments

Reasons for surface coating/treatment		Examples/comments
To prevent direct deterioration	Chemical attack	Attack by aggressive chemicals such as acids, sulphates, sugars and fertilisers
	Physical effects	Deterioration due to erosion/abrasion, salt crystallisation and freeze–thaw action (surface scaling or deeper disintegration)
To prevent indirect deterioration due to reinforcement corrosion	Loss of concrete alkalinity and steel passivation due to the ingress of acidic gases	Carbonation
	Premature initiation of corrosion due to ionic ingress	Ingress of chlorides in coastal environments or from de-icing salts
To limit or control ingress/contact	Waterproofing	Barriers to liquid water that can vary widely in their resistance to moisture vapour transmission; some systems are approved for contact with potable water
	Vapour/gas barriers	Barriers to moisture vapour, methane, radon and acidic gases, e.g. CO_2, SO_2 (NO)x
	Ease of cleaning and decontamination	Floors, walls in food processing areas, hospitals and nuclear installations
To enhance/maintain appearance	Colour and texture	Building facades
	Reflectance	Road tunnels and car parks
	Prevention of mould growth and dirt staining	Walls and floors
	Anti-graffiti coating/treatment	Assisting removal of graffiti
	Uniformity after repair	Overcoating following patch repairs
To enhance safety	Anti-slip/skid	Used with a scatter of fine aggregate on floors/roads
	Anti-static/electrically conductive systems	Floor coatings in manufacturing areas
	Road/floor markings	Defining specific areas by colour

Further details are given in Table 17.2.

These materials mainly involve keeping aggressive materials and chemicals out of concrete. Basically there are pore liners (H) that repel water (e.g. silane and siloxane), pore blockers (I) that soak into the surface and seal the porosity (e.g. resins) and conventional coatings (C) that adhere to the surface of the concrete.

Within BS EN 1504 Part 2 there are a large number of Standards that are used to characterise the coating system, many of which also govern factory control systems – ensuring that products are manufactured consistently. These are termed identification requirements (the material complies with its description) and performance requirements (the material performs according to the claims made for it). The manufacturer's quality control regime aims to test a representative number of samples for compliance. These tests are,

Table 17.2 Principles and methods according to EN 1504-9: 2008

Principle no. and definition	*Methods based on the principle*
1. Protection against ingress Reducing/preventing the ingress of adverse agents.	Pore-blocking impregnation* Surface coating with or without the ability to accommodate crack formation and movement
2. Moisture control Adjusting and maintaining the moisture content in the concrete within a specified range of values.	Hydrophobic impregnation* Surface coating
3. Concrete restoration	–
4. Structural strengthening	–
5. Physical resistance Increasing resistance to physical/mechanical attack.	Overlays or coatings Pore-blocking impregnation
6. Resistance to chemicals Increasing resistance of the concrete surface to deterioration by chemical attack.	Overlays or coatings Pore-blocking impregnation
7. Preserving or restoring passivity	–
8. Increasing resistivity Increasing the electrical resistivity of the concrete.	Limiting moisture content by surface treatments or coatings
9. Cathodic control Creating conditions in which potentially cathodic areas of reinforcement are unable to drive an anodic reaction.	Limiting oxygen content (at the cathode) by saturation or surface coating
10. Cathodic protection	–
11. Control of anodic areas	–

* The distinction between pore-blocking and hydrophobic impregnation is discussed in Section 17.4.

however, laboratory procedures, i.e. they would not be appropriate for QA purposes on site. BS EN1504-10 (BSI, 2003) covers Quality Control of the works and includes testing that can be carried out on site.

Whilst manufacturers usually take responsibility for many of these Standard tests, there are a number of requirements that must be noted by specifiers, exemplified as follows.

- For hydrophobic impregnations, there are two classes that relate to the depth to which the material penetrates the concrete on standard test blocks: Class I is < 10 mm, and Class II is > 10 mm. Similarly, there are two classes for drying rate coefficients. There is also a performance characteristic for resistance to diffusion of chloride ions.
- For pore-blocking impregnations, there are three classes of permeability to water vapour (Class I, permeable; Class II, medium permeability and Class III, dense against water vapour). Similarly, for impact loading there are three classes: Class I is the lowest, Class III is the highest impact resistance. Also, there are three classes for slip and skid resistance, dependent on the exposure (inside wet surfaces, inside dry and outside), although these classes may be modified to meet national regulations.
- For coatings, there are two classes for withstanding abrasion by traffic. Three classes for water vapour are also present, similar to impregnations. Thermal compatibility is divided into trafficked or untrafficked conditions for adhesion after various exposure cycles, with further subdivision into flexible, crack-bridging or rigid systems. For crack-bridging systems, the required crack accommodation should be selected by the designer with respect to local conditions, with *no* failures allowed. The impact resistance is again split into three categories, and there are two classes for antistatic coatings dependent on environment.

The final sections of BS EN 1504-2 consist of informative annexes:

- Annex A gives an example of the minimum frequencies of manufacturer's testing.
- Annex B gives useful examples of what designers need to specify for three separate cases.
- Annex C relates to the release of dangerous substances.
- Annex Z, which is over 30% of the document, relates to the Construction Products Directive and certification of conformity.

As with other parts of BS EN 1504, conformance to this part of the Standard does not guarantee that applying a surface protection system will provide the required level of performance. Even if the chosen system is the correct type, it does not guarantee that the material has been properly applied on site (appropriate and very useful site-tests are given in BS EN 1504-10).

Although the guarantee of a successful repair will still not be achieved, if a system has been properly designed, specified and applied it has the best chance of achieving the required end result (Atkins, 2007).

17.3 The specification of surface coatings/treatments

The use of protective surface barriers can be related to the age and condition of the concrete, as discussed below.

(i) New concrete, satisfactory quality: 'normal specification' of surface treatment

Surface barriers would normally be specified at the design stage where there is an obvious incompatibility between the performance of concrete and a particular service environment or demand. Hence, the predominant reasons for use are likely to be the prevention of direct deterioration and the control of ingress/contact.

In exceptionally aggressive environments, they may be used to enhance the resistance to indirect deterioration due to reinforcement corrosion. However, as surface treatments inevitably require maintenance, other options may be preferable (e.g., improved concrete mix design, increased cover, alternative forms of reinforcement).

Coatings may also be specified initially where both chlorides and sulphates are present (even for the lower classes of sulphate exposure) as sulphate-resisting Portland cement has a relatively low resistance to chloride ingress, due to its reduced tricalcium aluminate content. However, blended cements, such as Portland-Pfa or Portland-Blastfurnace cements, are usually specified to achieve the required durability in these conditions.

(ii) New concrete, unsatisfactory quality: 'remedial specification' of a surface treatment

Here, surface barriers may be used to alleviate potential deterioration due to reinforcement corrosion arising from an inadequate mix design, insufficient cover or the poor compaction/cure of concrete. It is well established that proprietary products can provide effective chloride and carbonation barriers over lengthy periods of service (although they are not a substitute for correct initial specification and placement of the concrete).

(iii) Concrete undergoing deterioration: 'repair specification' of a surface treatment

Where direct deterioration due to chemical/physical effects has occurred, reinstatement by the simple application of a surface treatment may be feasible at an early stage, particularly if a thick lining or render is used. It is more likely, however, that significant preparation will be required before coating (i.e. cutting back, cleaning and profile restoration).

For structures that have been in service for a significant period, a distinction must be made between the potential effectiveness of surface treatments applied before/after the initiation of reinforcement corrosion; also between chloride ingress and carbonation.

Prior to initiation

Appropriate surface treatments can effectively inhibit the progress of carbonation, reducing its rate to a negligible level. Hence, the service life of a structure can be extended considerably where carbonated concrete in the cover zone has not reached the reinforcement.

The value of surface treatments is more debatable when chloride ingress has occurred. The rate of chloride migration to the reinforcement may be reduced by preventing additional penetration and by limiting the internal moisture state of the concrete. However, it is unlikely that corrosion will be prevented in the longer term – or even the short/medium term if chloride ingress is at a high level prior to coating.

After initiation (following carbonation or chloride ingress)

In this case, surface barriers can only be effective by limiting the ingress of oxygen and/or moisture. As only very small amounts of oxygen are required to fuel the corrosion process, even the most (oxygen) impermeable surface treatment is unlikely to be effective. However, attempting to maintain the concrete in either a very dry or a water-saturated condition may assist in reducing the corrosion rate.

After initiation, therefore, surface treatments are likely to be cost-effective only where a relatively short extension of service life is required, or when used in conjunction with other methods such as sacrificial anode cathodic protection to reduce the work required from such systems and extend their life.

Prior to carrying out any repair, it is essential that a proper assessment is made of the defects in the concrete and their causes, and of the ability of the structure to perform its function.

17.3.1 Additional comments

Although a detailed discussion of the applications summarised in Tables 17.1 and 17.2 cannot be given here, concise information, in most cases, can be obtained readily from the publications listed in the Bibliography. Some brief comments regarding several particular uses are given below, however, as published information in these areas is more scattered. These examples also serve to illustrate the more general point that specialist knowledge is often essential, even when considering seemingly simple applications.

(i) Alkali–silica reaction (ASR)

The hypothesis has often been advanced that above-ground structures subject to ASR can be protected by surface waterproofing, thus maintaining the concrete in a sufficiently dry state to inhibit or arrest deterioration. The results of various laboratory studies seemingly support this hypothesis. However, there appears to be little, if any, documented evidence for the effectiveness of this approach on real structures. Field trials, and more general considerations, have also led to the conclusion that there are serious practical difficulties in attaining the ideal conditions within the concrete for the inhibition of ASR.

These difficulties may be influenced by a variable micro-climate in the vicinity of the affected concrete and by moisture ingress from other parts of the structure. Furthermore, many coatings and related treatments that provide excellent barriers against liquid water ingress have a relatively low resistance to moisture vapour transmission. Consequently, for this form of deterioration in particular, the simple extrapolation of laboratory results to in-service conditions is likely to be very speculative.

(ii) Waterproofing and carbonation

Concerns have arisen over the use of pore-lining penetrants, such as silanes, that eliminate periods of high internal moisture content while offering little, if any, protection against carbonation. However, the view that carbonation may be encouraged by the maintenance of an optimum internal moisture state (50–70% RH) does not appear to be supported by experience from real structures.

(iii) Control of reinforcement corrosion by the limitation of oxygen ingress

Oxygen is required both to sustain the electrochemical corrosion reaction at the steel reinforcement and also for the formation of the expansive rust that causes cracking and spalling of the concrete cover zone.

The potential use of surface coatings to control reinforcement corrosion by limiting oxygen ingress is noted in Table 17.2 (Principle 9 of BS EN 1504-9: 2008). However, in general, this approach is most unlikely to be effective because, as noted earlier, only very small levels of oxygen are required to fuel the corrosion process, while even the most impermeable coatings are not completely resistant to oxygen transmission. Furthermore, without total encapsulation, oxygen is likely to reach the reinforcement via other pathways.

17.4 Materials for surface coating/treatment

Many proprietary products are available for the surface treatment of concrete. They can be classified conveniently according to the main generic component, as shown in Figure 17.1.

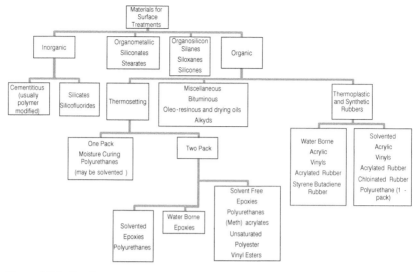

Figure 17.1 Coating types

An alternative classification scheme, based on the following categories, is also useful as it can be related more easily to particular applications. This approach has been adopted in European Standards BS EN1504-2 and 9 (BSI, 2004, 2008).

i   Materials that impregnate the concrete and that do not depend on the formation of a significant surface film. These may function in two ways:
  • Penetrants that line the pores with a water-repellent/hydrophobic layer, relying on surface tension effects to prevent the ingress of liquids by absorption (capillary action).
  • Sealers that fill and block the pores, thus offering some resistance to the ingress of liquids under a pressure gradient. A distinct surface film can be formed if the application rate is sufficient (usually > 2 or 3 coats) or if the concrete is particularly dense and impermeable.
ii Systems that depend on the formation of a continuous layer, thus shielding the concrete surface completely. Here, the following sub-division is useful:
  • Coatings, including high-build systems, that have a dry film thickness up to approximately 1–2 mm.
  • Heavy-duty linings/renders with a greater thickness.

This scheme is used in Tables 17.3, 17.4 and 17.5 where brief comments are given on the main performance characteristics of the more common systems (thick linings and renders have been excluded, to avoid repetition, as they are often based on similar generic components to those used in coatings).

Table 17.3 Classification and performance characteristics of hydrophobic penetrants/
pore-liners (water-repellents)

Common generic types	Main performance characteristics
Silanes **Siloxanes** **Silicones** Numerous variations in chemical type, content of active material, and solvent system are available May be 100% active Silanes/siloxanes react in situ with the concrete; silicones generally undergo drying only. In both cases, the pores/ capillaries are lined with a hydrophobic layer	Protect against water/chloride ingress in the absence of hydrostatic pressure (questionable on ponded surfaces) Some systems are approved for use on UK highway structures Negligible effect on ingress of moisture vapour, carbon dioxide or oxygen No benefit for resistance to chemicals, abrasion or impact Negligible effect on concrete appearance Durability generally expected to be good, more so for silanes/siloxanes, but claims vary Relatively easy to apply/re-apply with minimal surface preparation Substantially dry concrete required Low temperatures acceptable but hot/windy conditions unfavourable (use of a 'paste' form preferred)

Table 17.4 Classification and performance characteristics of pore-blocking sealers

Common generic types	Main performance characteristics
Generally solvented, although some water-based dispersions can be suitable **Thermoplastic** **(acrylics, pu's)** Physically drying only **Thermosetting** **(epoxy, pu's)** Chemically reactive and physically drying May be one- or two-pack systems Wide variations in solids content, solvent/polymer type are possible	Enhanced resistance to ingress of liquids, gases, vapours, chemical attack and abrasion Effectiveness depends on severity of service conditions, specific requirements and porosity of concrete/number of applications (2–3 is usual) Some slight darkening of the concrete is usual and the formation of a surface glaze may occur Generally have reasonably good long-term effectiveness as there is no dependence on surface adhesion Relatively easy to apply/re-apply with minimal surface preparation Can also be used effectively as a primer/sealer prior to coating on porous/friable surfaces

Table 17.5 Performance characteristics of different coating systems

Common generic types	Main performance characteristics
Cementitious, polymer-modified Generally ~ 2 mm thick (2 coats) Supplied as two-pack systems (blended cement/fine fillers and aqueous polymer dispersion) or as one-pack (blended cement/filler /redispersible powdered polymer that is mixed with a specified quantity of water)	Generally provide excellent barrier properties (water, chlorides, carbon dioxide, oxygen) Relatively low resistance to moisture vapour transmission Generally unsuitable for exposure to strong acids but can provide some resistance to sulphate attack Can possess a useful degree of flexibility Good durability under a wide range of service conditions (including immersed and below ground) Reasonably tolerant of various surface/application conditions; temperature extremes may require special precautions or limit use Acceptable maintenance requirements
Silicates Filled, pigmented systems (much simpler products are used as low-cost dust-proofers) Reaction with the concrete surface assists adhesion	Excellent durability as breathable and weatherproof decorative coatings Some anti-carbonation properties Resistant to some chemicals Reasonably tolerant of various surface and application conditions Easy to maintain
Solvent and water-borne thermoplastics and elastomers Physically drying Formulations and performance vary widely	Can provide excellent barrier properties (water, chlorides, carbon dioxide) and good breathability Some systems suitable for exposure to aggressive chemicals Good durability and retention of appearance Generally not suitable for immersed/below ground applications or demanding abrasion resistance Can maintain a barrier when applied to some cracked concrete (fibre/fabric reinforcement may be required) Can be applied under widely varying conditions with relative ease (particularly when solvent-borne) Good surface preparation required – often need a fairing/levelling coat and/or primer Relatively easy to maintain

Common generic types	Main performance characteristics
Thermosetting systems Solvented, water-borne or solvent-free Undergo chemical reaction during cure Formulations and performance vary widely Extremely versatile – can be 'tailored' to specific requirements Usually two-pack Solvent-free systems are particularly suited to very high-build applications (up to at least 250 µm/coat)	Excellent barrier properties against liquids, gases and moisture vapour Acceptable appearance although chalking and colour instability prevalent with some products Excellent durability under all conditions including frequent contact with, or immersion in, aggressive chemicals Good impact/abrasion resistance May accommodate cracks in the concrete (likely to require fibre reinforcement) Generally require a high level of surface preparation for potential to be fully realised Systems available for application to wet surfaces (or, in some cases, under water) Use may be limited/not possible at low temperatures (generally below 5°C) Cure rate often slow, even at 'normal' temperatures Fairly demanding application requirements (specialist applicator usually advisable) Maintenance often difficult due to potentially poor old/new intercoat adhesion
Bituminous systems Hot-applied (generally with fabric reinforcement) Cold-applied Water-borne emulsions or solvent-borne (± reinforcement)	Used mainly for waterproofing applications below and above ground Reasonably durable but can tend to embrittle with age, particularly when exposed to sunlight and hot conditions Relatively easy to apply/re-apply with good tolerance of surface and ambient conditions

It should not be assumed that all proprietary products will conform to this overview, as many variations are possible; in addition, a combination of generic types may be used to provide a sealer/primer/intermediate and top-coat system. Generic characteristics (e.g. 'all epoxies bond well and have good chemical resistance') can also encompass significant variations, depending upon the detailed composition (formulation) of a product.

Properties are not discussed here in detail, but a summary of characteristics that may have to be considered for various applications is given in Table 17.6. Where corresponding information is not provided on data sheets, suppliers should be consulted (on occasion, specific tests may need to be commissioned).

17.5 Surface preparation and application

Concrete is inherently suitable as a substrate for coatings, although it can present problems that are not met with steel, for example:

- The (uncarbonated) surface has a high alkalinity.
- Concrete surfaces are invariably rougher and often have partly open air-voids ('blow-holes') and protrusions (e.g., fins/nibs or grout runs).
- Concrete is absorbent to a varying degree and frequently has a relatively high moisture content.
- Surfaces may be dusty and friable and often have a thin, relatively weak surface layer (laitance).
- Contamination due to mould release agents, curing membranes and fungal growth is common (note: curing membranes designed to degrade may persist in the absence of sunlight; some specific membrane types can be overcoated).

For a coating to perform satisfactorily, application to a clean, sound surface is normally essential, thus enabling the development and retention of the maximum bond strength. Certain demands may be relaxed somewhat for many penetrants and sealers as surface adhesion is less critical, although some preparatory work may still be necessary; contaminants that hinder penetration, for example, should be removed.

The need to remove an existing coating can vary, depending upon its condition, including adhesion, and its chemical/physical/adhesive compatibility with the new system. Identification of the existing coating will assist significantly in determining the best course of action; in-situ testing, e.g. for adhesion, can also be advisable. Specific performance properties required of the original system should be identified and maintained, where necessary, by the new coating.

For new concrete, controlled permeability formwork can simplify surface preparation as it provides a sound surface with minimal blow-holes and also eliminates contamination by shutter release agents (McKenna, 1995).

Table 17.6 Properties of surface coatings/treatments required by the specifier/applicator

Unmixed and freshly mixed	Transition to the dry/cured state	Fully cured Short term	Long term
Shelf life	Effects of ambient/ substrate temperature	Dry film thickness	Change of short-term properties on exposure to service conditions
Storage requirements, particularly temperature	Sensitivity to ambient/ substrate moisture, %RH	Adhesion	
Flash point	Usable (pot) life	Colour, texture, hiding power, gloss and reflectance	Accommodation of crack formation and movement
Volatile components health, safety and environmental considerations	Gel time	Barrier properties	Abrasion resistance
Taint, e.g., of nearby foodstuffs	Reaction exotherm (for some thicker systems)	Mechanical properties	Effects of thermal cycling/shock
Density and coverage rate	Rate and extent of cure/ drying vs time	Fire performance	
Need for priming	Cure shrinkage (only for certain thick systems)	Electrical properties	Resistance to water, chemicals, biological attack/mould growth, radiation
Mixing requirements	Over-coating interval	Slip/skid resistance	
Application properties/methods		Effect on potable water	
		Ease of cleaning and nuclear decontamination	Resistance to weathering
		Resistance to graffiti and the ease of its removal	Cleanability
			Ease of maintenance

The more common methods of surface preparation and coating application are outlined in the following sections, where it is assumed that the surface is basically sound. Hence, the following processes are not discussed: patch repair/re-casting to rectify reinforcement corrosion, spalling or mechanical damage; the treatment of joints and cracks; and the stemming of water ingress/seepage.

Where a coating does form part of a complete repair package – for example, following deterioration due to reinforcement corrosion – many methods of in-situ assessment are used routinely. However, these techniques contribute to surveys aimed at deriving a repair strategy and specification rather than solely to quality control procedures. They include the determination of concrete strength, depth of cover, corrosion activity and permeability and are discussed in detail in several publications: Concrete Society TR54 (Concrete Society, 2000), the book 'Testing of Concrete in Structures' (Bungey et al., 2006) and CIRIA Technical Note 143 (CIRIA, 1992).

Health, safety and environmental issues are not discussed here, but it must be noted that certain assessments and actions are covered by statutory regulations. Guidance can be obtained from published sources (see, for example, BS EN1504-10 (BSI, 2003) and references therein) and from product suppliers (Health & Safety Data Sheets with a regulated content must be made available).

17.5.1 Surface preparation

Cleaning

Many techniques are used for preparatory cleaning and the most appropriate will depend upon individual circumstances. The following methods are given in approximate order of increasing severity:

- Wet scrubbing with emulsifying detergents and, where necessary, biocides
- Low-pressure water cleaning
- Steam cleaning
- Acid etching
- Hand/power wire brushing
- Grinding
- Wet or dry abrasive blasting (with vacuum recovery)
- Shotblasting
- Scarifying/planing (cutting teeth on rotating discs)
- Needle gunning
- Scabbling
- (Ultra) high-pressure water jetting (100–300 MPa)
- Flame blasting
- Milling (cutting teeth on a rotating drum).

Further advice on cleaning methods is given in several standards and other publications (CRA, 1992; Dept of Transport BD43/90, 1990, Dept of Transport BA33/90, 1990 and Dept of Transport BD27/86, 1986). Additionally, a range of plaques has been produced that replicate the typical profiles given by various methods of preparation. These specimens are also linked to the use of coatings/treatments of different thickness, see the ACI Concrete Repair Manual (ACI, 1999). However, it is recommended here that this correlation should be used for general guidance, rather than prescriptively, as supplier's recommendations may differ and should take precedence.

Relatively simple washing techniques may not be effective where contaminants such as oil and grease have been absorbed by the concrete. They will also be ineffective if the contamination is merely spread further. Following wet cleaning, thorough washing with fresh water should be carried out, as some detergent and biocide residues can have an adverse effect on adhesion. Similar problems can arise if chemical strippers, used to remove old coatings, are allowed to dry – washing off all stripping residues can then become difficult.

Readily available proprietary products that contain biologically active micro-organisms can rapidly degrade hydrocarbon residues and greatly assist removal.

Acid etching is now used infrequently, due mainly to the associated health and safety risks. Additionally, the absorption, via porous areas/cracks, or surface retention of certain ionic species, e.g. chlorides, can be detrimental.

Wet grit-blasting and vacuum dry-blasting are commonly favoured because they are effective techniques that allow good control of the cleaning process, while generally having no significantly detrimental effects on the surface. Health, safety and environmental risks are also reduced, compared with 'open dry-blasting', although the wet process may create containment and disposal problems. Many typical concrete surfaces require no more than very light blasting ('sweep-blasting') prior to coating.

High-pressure water jetting is an effective method for dealing with many contaminated or unsound surfaces. Equipment is available that allows good control of the process, giving an acceptable profile, and recycling of the jetting water.

Power driven mechanical methods, such as scarifying, needle gunning and scabbling, are widely used and can be very effective for the removal of defective surface layers and firmly adhering or deeply absorbed contaminants. However, they can also be too aggressive, producing micro-cracking of the aggregate, which may be detrimental to bond strength, and an overly deep texture. Excessive power wire-brushing can lead to undesirable polishing of the surface.

Milling, which is also a very aggressive method, can remove an appreciable depth of concrete, but the large, heavy equipment is only suited to horizontal surfaces.

Flame blasting is used infrequently. It fractures and removes the concrete surface by superheating the pore water, thus generating expansive forces.

It is essential that the surface is thoroughly washed with clean water and/ or vacuumed following preparation, thus ensuring that all loose material is fully removed. A simple 'water droplet' test may also be used to provide an indication of surface cleanliness – contamination will tend to prevent spreading and absorption of the water.

Void filling and levelling

Durability (for other than permanently dry conditions) and aesthetics require that coatings form a continuous film, free of pinholes and with an adequate thickness over the entire surface. Consequently, any significant unevenness must be dealt with following initial preparation.

With some thicker products, this may be achieved by direct application to a prepared surface. In other cases, however, small voids and 'blow-holes' (which usually become more exposed and enlarged during initial preparation) must be filled and the surface must be levelled prior to coating. Filling and re-profiling is carried out by 'bagging-in', by use of a 'scrape/skim coat' or by application of a (typically) 1–2 mm thick 'fairing coat'.

Cementitious slurries and proprietary products, generally polymer-modified, are most commonly used, although pastes or fine mortars based on a reactive resin, e.g. an epoxy, can also be useful. Compatibility with the coating system and any other specific requirements should be checked with the supplier.

17.5.2 Application

It must always be ensured that the coating/treatment is suitable for use on an alkaline concrete surface – some products are readily degraded and debonded by alkalies, particularly in moist conditions.

Excessively absorbent or weak and friable/dusty surfaces may be brought to a satisfactory condition with a sealer/stabiliser. The over-application of some products must be avoided and, in general, the sealer and other components of the system should be obtained from the same supplier to ensure compatibility.

The required age of new concrete prior to coating varies, but a minimum of 21–28 days under reasonable ambient conditions is not uncommon. Normally, this allows for both early shrinkage effects and drying. Particular attention must be given to new floor slabs and screeds, as very long periods can be required to attain a moisture content that will not cause premature failure of a coating (FerFa, 2000). Specific products – surface damp-proof membranes – are available for circumventing this problem.

More generally, for mature concrete elements, the tolerable moisture level varies according to the particular coating system – its ability to displace water, the bonding characteristics and the permeability to moisture vapour are particularly relevant properties – and the supplier's recommendations should always be observed. Coatings are available for application to very wet surfaces, even under water. However, this capability is formulation dependent rather than being typical of a generic type. For the adequate absorption of penetrants and sealers, water-free pores/capillaries are obviously essential.

The most appropriate method of application depends on the product type/viscosity, the area to be treated and its continuity/accessibility. Common methods are:

- Spraying
- Rolling
- Brushing
- Flood coating.

Airless spraying, which utilises a high-pressure pump to atomise the material, is most suitable for the rapid application of relatively low viscosity products to unobstructed large areas. Air-spraying is suited to more viscous products that are difficult to atomise – in this case, masking of the surroundings is likely to be necessary, due to the overspray mist that is created. Both methods become difficult to use in windy conditions.

Air-assisted spraying, using plant similar to that employed for 'guniting', can place very thick products containing significant amounts of sand, large filler particles or fibres. A finishing operation, typically trowelling, is required and wastage, due to rebound, is inevitable.

Application by roller is commonly used for areas of intermediate size, i.e. those that are too small/large for spraying/brushing, respectively. Brush application is most suited to relatively small and/or inaccessible areas, and to products, usually primers, that must be worked vigorously into the surface.

Flood coating is a technique that utilises a low-pressure spray to saturate a surface uniformly with very low viscosity penetrants, particularly silanes/siloxanes. A free-running wet front is maintained, working from top to bottom on a vertical surface. A related technique, used on floors, employs a 'squeegee' to apply and spread flood coatings of penetrating sealers.

The range of ambient temperature and humidity that is acceptable during application and drying/curing varies, according to both generic type and formulation, and recommendations should always be available from the supplier. In general, it is good practice to store materials in a controlled temperature environment for at least 24 hours before use.

Particular care is required when temperatures are 'low and falling' as condensation can affect both adhesion and appearance. It can become essential to monitor ambient conditions and surface temperature to ensure that application is not carried out in proximity to the dew point.

For reactive coatings, epoxies in particular, the recommended maximum period between successive coats must be observed (with allowance being made for on-site temperatures that differ significantly from those given on the data sheet). An excessively long delay between coats can lead to poor intercoat adhesion because the surface of the initial coat becomes less receptive to bonding as cure progresses.

The period recommended for the attainment of full cure/drying, which can be as long as 10–14 days, must be allowed prior to imposing aggressive service conditions, for example: chemical contact, immersion in water or heavy trafficking. Where a data sheet refers only to a single cure temperature, such as 20°C, a significantly longer period is likely to be required under colder, more variable site conditions.

17.5.3   *Quality control*

The many coating variations available differ in their tolerance of surface and ambient conditions. Consequently, an appropriate process for coating application must be specified and followed for any given set of circumstances. Good workmanship, supervision and inspection are essential. A variety of proprietary instruments provide applicators, supervisors and inspectors with objective means of assessing the quality of coating work on concrete.

Inspection and supervision should ensure that surfaces are well prepared and suitable for application of the coating/treatment. In-situ strength measurement (e.g. by pull-off testing) may be used to ensure that the substrate is mechanically sound and receptive to bonding. The principle of the test is simply to apply a tensile load to a circular metal dolly, bonded to the surface, until failure occurs. The test has been very widely used for many years and is described in numerous national standards and elsewhere (EN 1504-2, EN1504-10 and BS EN 1542: 1999, 'Products and Systems for the Protection and Repair of Concrete Structures, Test Methods: Measurement of Bond Strength by Pull-Off').

Various designs of pull-off adhesion tests are available from many equipment suppliers. The dollies, usually of steel or aluminium, are commonly bonded to the surface using a rapid-curing two-pack epoxy resin, allowing a same/next day test. Because most concrete has coarse aggregate, it is generally advisable to use dollies with a minimum diameter of 50 mm.

Determination of the moisture level in the concrete is often advisable; in the case of floor slabs/screeds, this measurement is usually required by specification (see, for example, the application guide for resin flooring (FerFa, 2000)).

Particular attention should be given to the limitations of surface measurements: low moisture levels may not reflect the content at depth. While initial adhesion of a coating may be quite satisfactory, failure may occur once the relatively impermeable covering allows a high moisture level to be established at the bonded interface.

On floor slabs/screeds, this potential problem is avoided by stipulating in various codes that moisture content is determined only by the 'hygrometer

/ sealed box' method – essentially determination of the equilibrium relative humidity in an enclosed air space formed above the concrete (see BS 8204-6, 'Screeds, Bases and In Situ Floorings, Part 6: Synthetic Resin Floorings – Code of Practice' and BS 8203, 'Code of Practice for Installation of Resilient Floor Coverings'). Equipment developed for this test is available commercially.

Application of the correct coating thickness is normally essential for satisfactory appearance and durability. It can be monitored approximately by recording the consumption rate over known areas – with due allowance for the effects of surface roughness and wastage. Wet film gauges can be useful for carrying out rapid spot-checks during application, but their accuracy is limited unless the surface is reasonably smooth.

Dry film thickness is more difficult to check on concrete than on metallic substrates, although direct reading instruments based on ultrasonic methods are now available. Monitoring may be carried out via metallic coupons attached to the surface prior to coating, thus allowing use of the more common electronic thickness gauges.

Depending on the substrate/coating texture, dry film thickness can also be measured by cutting or boring through the coating using a tool whose tip has a defined angular geometry (for example, the very well-known paint inspection gauge). The total film thickness or that of individual layers is then read off from a microscope graticule scale. Instruments using a borer are generally better suited than the linear cutters when harder, more brittle coatings are examined. Figure 17.3 shows a typical dry film thickness gauge.

In some circumstances, it may be necessary to take cored specimens so that the coating thickness along sawn edges can be determined with a laboratory microscope. Core specimens can also be used to determine the penetration depth of hydrophobic impregnants and pore-blocking sealers.

Most coating systems require the application of at least two coats and, in some cases, different colours can assist in achieving the correct build-up of successive layers.

Once the coating has fully dried/cured, adhesion can be checked by pull-off tests using one of the many instruments available commercially, or by a cross hatch cutter (Figure 17.4). It is often recommended or specified that the test area is isolated by coring through the coating and into the concrete. This provides a well-defined test area and eliminates restraint from the coating adjacent to the dolly.

Procedural details, the extent of testing and performance criteria should be agreed beforehand. It is generally advisable to agree to test locations, the level of replication, and performance criteria at the outset. The simplest requirement is that failure should occur only (or predominantly) in a cohesive mode, in either the concrete or the coating, and not adhesively at the substrate or between coats. A required failure stress is usually expressed as an average together with a minimum for any one result.

Figure 17.2 Wet film thickness gauge to assess adequacy of coating immediately after application (courtesy Paint Test Equipment).

Figure 17.3 Dry film thickness gauge and cross hatch cutter (courtesy Elcometer Ltd).

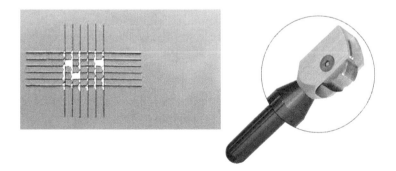

Figure 17.4 Cross hatch cutter to assess coating adhesion.

Other in-situ properties of coatings on concrete that can be checked easily with portable instruments include the following:

- the presence of pin-holes ('holidays') using a high voltage detector
- the slip (or skid) resistance of horizontal trafficked surfaces
- aspects of appearance, such as gloss, colour, and light-reflectance values.

Trial areas can be useful, particularly on large contracts and where maintenance of a high level of performance is critical. By this means, standards of workmanship can be agreed – and any problems resolved – at an early stage. Where necessary, extensive in-situ testing, or core removal for laboratory assessment, can also be carried out conveniently at this stage.

Common problems that occur on site are variable, inadequate coating thickness (can be checked by wet film thickness gauge, Figure 17.2, and dry film laboratory tests on cores) and using an incorrect material (one material specified, another one used!). The latter can be resolved by careful supervision on site and techniques are available to positively identify the coating in the event of a dispute (infra-red spectroscopy and electron microprobe analysis of the pigment to identify elements present).

References

ACI (1999) *Concrete Repair Manual*, pp. 619–661, International Concrete Repair Institute, ACI International.

Atkins, C. (2007) 'EN1504 Part 2, Surface Protection Systems.' *Concrete*, September, pp 13–14.

BS EN 1504-2: 2004 'Products and Systems for the Protection and Repair of Concrete structures. Definitions, Requirements, Quality Control and Evaluation of Conformity. Surface Protection Systems for Concrete.'

BS EN 1504-9: 2008. 'Products and Systems for the Protection and Repair of Concrete Structures. Definitions, Requirements, Quality Control and Evaluation of Conformity. Part 9: General Principles for the Use of Products and Systems.'

BS EN 1504-10: 2003 'Products and Systems for the Protection and Repair of Concrete Structures. Definitions, Requirements, Quality Control and Evaluation of Conformity. Site Application of Products and Systems and Quality Control of the Works.'

Bungey, J. H., Millard, S. G. and Grantham, M. G. (2006) *Testing of Concrete in Structures*, fourth edition, London: Routledge.

CIRIA (1992) Technical Note 143, London: CIRIA

Concrete Repair Association (UK) (1992), *The Application and Measurement of Protective Coatings for Concrete – Guidance Note*, Borden: Concrete Repair Association.

Concrete Society (2000) *Diagnosis of Deterioration in Concrete Structures*, Technical Report No. 54, London: Concrete Society.

Department of Transport/Highways Agency (UK) (1986) 'Materials for the Repair of Concrete Highway Structures', Departmental Standard BD 27/86, London: Department of Transport.

Department of Transport/Highways Agency (UK) (2003) 'Impregnation of reinforced and prestressed Concrete Highway Structures using hydrophobic pore-lining impregnants' Departmental Advice Note BD 43/03, London: Department of Transport.

FerFA (2000, revised 2009) *FeRFA Guide to the Specification and Application of Synthetic Resin Flooring*' Aldershot: The Resin Flooring Association. (The current version of this document can be downloaded from www.ferfa.org.uk)

McKenna, P. (1995) 'Controlled Permeability Formwork.' Current Practice Sheet 10, Concrete Bridge Development Group. Accessed July 2010 at http://www.cbdg.org.uk/pdfs/cps10.pdf

Price, W.F. (2000) *Controlled Permeability Formwork*, CIRIA Report C511, London: CIRIA and The Concrete Society.

Bibliography

Allen, R T L, Edwards, S C, and Shaw, J D N (1993) *The Repair of Concrete Structures*, 2nd edn, Blackie Academic and Professional

American Concrete Institute (1985) *A Guide for the Use of Waterproofing, Dampproofing, Protective and Decorative Barrier Systems for Concrete*, ACI 515. IR-79, Detroit, MI: ACI.

American Concrete Institute (1998) *ACI Manual of Concrete Practice*, Detroit, MI: ACI.

Bassi, R. and Roy, S.K. (2002) *Handbook of Coatings for Concrete*. Dunbeath: Whittles Publishing.

BS 6150: 1991. 'Code of Practice for Painting of Buildings'. BSI, 1991.

BS 8221-1: 2000. 'Code of Practice for Cleaning and Surface Repair of Buildings – Part 1: Cleaning of Natural Stones, Brick, Terracotta and Concrete'.

Concrete Society (1997) *Guide to Surface Treatments for Protection and Enhancement of Concrete*, Technical Report No. 50. London: The Concrete Society.

Department of Transport/Highways Agency (UK) (1990) 'Impregnation of Concrete Highway Structures', Departmental Advice Note BA 33/90, London: Department of Transport.

Dhir, R K and Green, J W (1990), *Protection of Concrete*, Proceedings of the International Conference held at the University of Dundee, Scotland, UK on 11–13 September 1990. London: E&FN Spon.

Doran, D K (1992), *Construction Materials Reference Book*, London: Butterworth-Heinemann.

Mailvaganam, N P (1991), *Repair and Protection of Concrete Structures*, Boca Raton, FL: CRC Press.

Mays, G (1992), *Durability of Concrete Structures : Investigation, Repair, Protection*, London: E&FN Spon.

Schmid, E V (1998) *Exterior Durability of Organic Coatings*, London: FMJ International Publications.

Index